Finding the Balance Between Schoolhouse and On-the-Job Training

Thomas Manacapilli, Alexis Bailey, Christopher Beighley, Bart Bennett, Aimee Bower

Prepared for the United States Air Force
Approved for public release; distribution unlimited

 PROJECT AIR FORCE

The research described in this report was sponsored by the United States Air Force under Contract F49642-01-C-0003 and FA7014-06-C-0001. Further information may be obtained from the Strategic Planning Division, Directorate of Plans, Hq USAF.

Library of Congress Cataloging-in-Publication Data

Finding the balance between schoolhouse and on-the-job training /
   Thomas Manacapilli ... [et al.].
     p. cm.
   Includes bibliographical references.
   ISBN 978-0-8330-4045-9 (pbk. : alk. paper)
   1. Aeronautics, Military—Study and teaching—United States. 2. United States.
Air Force. Air Education and Training Command—Evaluation.
I. Manacapilli, Thomas.

UG638.F55 2007
358.4'150973—dc22

2007011616

The RAND Corporation is a nonprofit research organization providing objective analysis and effective solutions that address the challenges facing the public and private sectors around the world. RAND's publications do not necessarily reflect the opinions of its research clients and sponsors.

RAND® is a registered trademark.

Published 2007 by the RAND Corporation
1776 Main Street, P.O. Box 2138, Santa Monica, CA 90407-2138
1200 South Hayes Street, Arlington, VA 22202-5050
4570 Fifth Avenue, Suite 600, Pittsburgh, PA 15213-2665
RAND URL: http://www.rand.org/
To order RAND documents or to obtain additional information, contact
Distribution Services: Telephone: (310) 451-7002
Fax: (310) 451-6915; Email: order@rand.org

Preface

This document addresses the trade-off between centralized initial skills training (IST) ("schoolhouse" training) and decentralized on-the-job training (OJT). The objective is to determine the most cost-effective combination of IST and OJT. Prior assessments of training have typically not fully evaluated how changes in IST affect the costs and other burdens of OJT.

In July 2003, the Air Force Council requested two training studies (later referred to as the AF Council–Directed Training Review).[1] The first study proposed a one-day reduction per year for three years in the course sequence for each Air Force specialty code (AFSC) in order to achieve identified fiscal year 2005 (FY05) Amended Program Objectives Memorandum (APOM) manpower and dollar adjustments. The second study was a comprehensive review of flight training, technical training, and education courses. Neither of these efforts addressed the productivity effect of adding or reducing specific pieces of course material or the change in OJT costs after graduates reach the field. Additionally, estimates of technical training costs rarely, if ever, include any consideration of the costs of OJT. Such costs include the limited productivity of trainees, the effect on supervisors, the effect of the supervisor-to-trainee ratio, and the disruption of training due to deployments.

The Vice Commander, Air Education and Training Command (AETC/CV), and RAND Project AIR FORCE (PAF) conceived

---

[1] The Air Force Council is chaired the by the Air Force Vice Chief of Staff. Members include the Air Staff Deputy Chiefs of Staff.

another approach to evaluating training reductions. Building on PAF work that used estimated effects on productivity to evaluate reenlistment bonuses, the research team identified a similar approach to finding a preferred mix of IST and OJT. A key part of this analysis was including both IST and OJT costs in determining the full cost of human capital development. This document describes a methodology for deriving these costs and applies the methodology to seven Air Force specialties (Aerial Cryptologic Linguist, AFSC 1A8X1; Cryptologic Linguist, AFSC 1N3XX; Tactical Aircraft Maintenance, F-15, AFSC 2A3X3A; Aerospace Maintenance, B-1/B-2, AFSC 2A5X1E; Special Purpose Vehicle Maintenance, AFSC 2T3X1; Fire Protection, AFSC 3E7X1; and Security Forces, AFSC 3POX1).

AETC/CV and the Air Force Deputy Chief of Staff for Manpower and Personnel (AF/DP) sponsored this research. The study was begun in the fall of 2004 as part of a project entitled "Cost and Productivity of Technical Training vs. On-the-Job Training Analysis." The work was conducted within PAF's Manpower, Personnel, and Training Program. This report should be of interest to leaders and staffs involved in planning and managing the Air Force's technical training pipeline. This includes elements of the Air Staff, as well as AETC organizations from Headquarters, Second Air Force, and training wings and squadrons down to the individual course level.

RAND Project AIR FORCE

RAND Project AIR FORCE (PAF), a division of the RAND Corporation, is the U.S. Air Force's federally funded research and development center for studies and analyses. PAF provides the Air Force with independent analyses of policy alternatives affecting the development, employment, combat readiness, and support of current and future aerospace forces. Research is conducted in four programs: Aerospace Force Development; Manpower, Personnel, and Training; Resource Management; and Strategy and Doctrine.

Additional information about PAF is available on our Web site at http://www.rand.org/paf.

Contents

Figures

Tables

Summary

How should enlisted initial training be divided between centralized initial skills training (IST) and decentralized on-the-job training (OJT)? This document provides recommendations to address this question based on a cost-benefit analysis of seven Air Force specialties. The underlying methodology includes a mechanism for developing specialty learning and productivity curves that are used to capture the full human capital development (HCD) cost (IST and OJT). All too often, only IST costs are considered when "pricing" training. When this is done, the overall cost to train an airman is seriously underestimated. When the full costs are considered, decisions related to the length of IST are better informed.

The Air Force typically trains 30,000 to 40,000 new airmen in some 300 specialties each year. We estimate IST costs at $700 million per year, with OJT costs reaching perhaps $1.4 billion each year. In developing new airmen to required levels of productivity, IST and OJT can, to some degree, substitute for each other. An appropriately designed cost-benefit analysis is necessary to find the best balance between them.

Productivity is difficult to measure. We equate a productive airman with one who possesses the skills to be fully mission capable. But then, in what ways must an airman in a particular specialty be skilled in order to be fully mission capable? How should a training program prepare airmen to obtain these skills and this experience? At what skill level do airmen currently graduate from IST? How does productivity increase with time in OJT?

To answer these questions, we fielded a survey among E-6s[1] and above from each of seven Air Force specialties[2] in four of the major commands (MAJCOMs)[3] in order to determine the effectiveness of the current IST and OJT training programs. A portion of the questions was designed to elicit the manner in which IST prepares airmen and how OJT increases productivity over time. We statistically averaged the responses and created the productivity curves shown in Figure S.1. Using these curves, we were able to decompose force costs into two parts: costs associated with productive effort (the proportion below the curve) and HCD costs (the proportion above the curve).

How does productivity change when IST course length changes? To make this assessment, we again relied on survey respondents to estimate the impact in both time and productivity. Unlike other attempts to determine the result of marginal (1-, 2-, or 3-day) changes in IST course length, we chose to request specific course material that could be added or deleted, the associated change in the course length, and the corresponding increase or decrease in productivity at graduation. For example, Figure S.2 shows curves statistically generated to fit the survey responses. Note that most of these curves increase initially but then level off, indicating that course lengths longer than a certain number of days will have diminishing impacts on productivity.

In our final methodological step, we combine the productivity curves with the course length changes and costs to examine total HCD costs. Costs include personnel cost, initial skills and advanced training costs, and other OJT-related costs, such as equipment and supervisor time. We use manpower data to determine the average number of

---

[1] Because of the small numbers in the airborne linguist (1A8X1) and cryptologic linguist (1N3XX) specialties, we also included E-5s from the 1N3XX specialty and E-4s and E-5s from the 1A8X1 specialty. The grades E-6 and above are primarily supervisors.

[2] Airborne Cryptologic Linguist (AFSC 1A8X1); Cryptologic Linguist (AFSC 1N3XX); Tactical Aircraft Maintenance, F-15 (AFSC 2A3X3A); Aerospace Maintenance, B-1/B-2 (AFSC 2A5X1E); Special Purpose Vehicle Maintenance (AFSC 2T3X1); Fire Protection (AFSC 3E7X1); and Security Forces (AFSC 3POX1).

[3] Air Combat Command, Air Mobility Command, Air Force Space Command, and Air Education and Training Command. Discussions with subject matter experts indicated that the selected MAJCOMs would provide an adequate sample of experiences for these AFSCs.

Figure S.1
Comparison of Fitted Productivity Curves

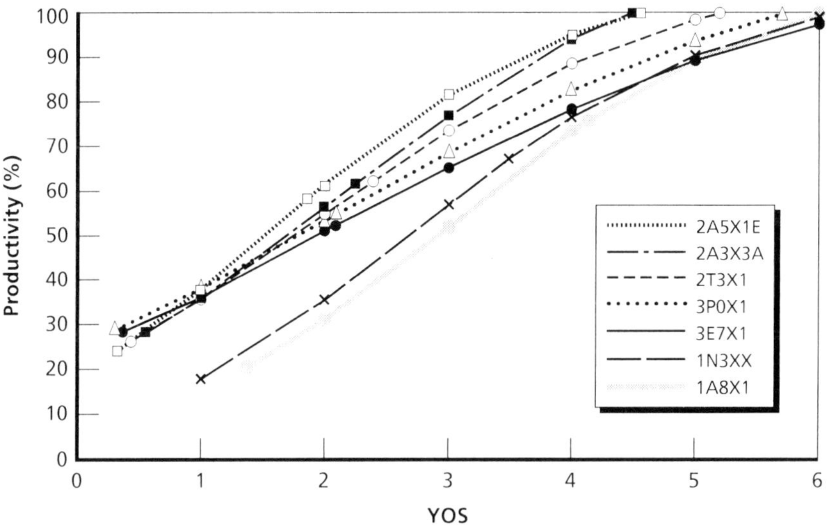

NOTE: 1A8X1 = Airborne Cryptologic Linguist; 1N3XX = Cryptologic Linguist;
2A3X3A = F-15 Maintenance; 2A5X1E = B-1/B-2 Maintenance; 2T3X1 = Special
Purpose Vehicle Maintenance; 3E7X1 = Fire Protection; 3POX1 = Security Forces.
RAND MG555-S.1

airmen in each year of service. Combining the data, we can examine
trade-offs among manpower, productivity, and cost.

Applying our analysis to the seven selected specialties, we achieved
the results described in Table S.1. Our analysis of the survey responses
was performed on a variety of dimensions. In addition to quantitative results, we included a qualitative analysis of comments from the
survey respondents. Column 3, "Write-in Comments," summarizes
free-response comments to questions suggesting course changes. The
"Add vs. Drop" column is a comparison of the number of responses to
specific questions on adding and deleting course content. The "Incremental Change Functions" column provides the direction of the average change in productivity for suggested changes to IST. The "Steady
State Analysis" column summarizes the results of the steady state cost-
benefit analysis.

Figure S.2
Changes in Productivity as a Function of Changes in Course Length

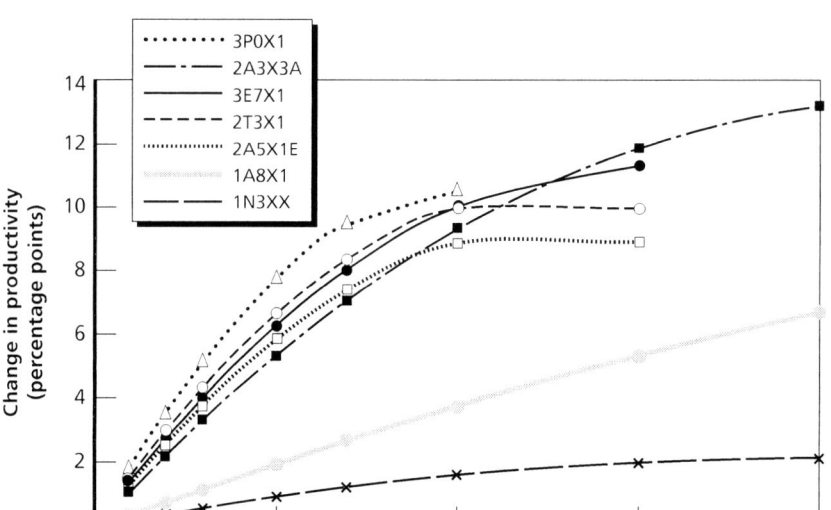

NOTE: 1A8X1 = Airborne Cryptologic Linguist; 1N3XX = Cryptologic Linguist;
2A3X3A = F-15 Maintenance; 2A5X1E = B-1/B-2 Maintenance; 2T3X1 = Special
Purpose Vehicle Maintenance; 3E7X1 = Fire Protection; 3POX1 = Security Forces.
RAND *MG555-S.2*

We recommend that the utilization and training workshops
(U&TWs) for each of the specialties perform a closer examination of
IST course length changes that appear to lower overall HCD costs
for the same or greater productivity. For AFSCs 1A8X1 and 1N3XX,
there is strong evidence that reductions in the course length would
result in only a small reduction in immediate productivity of graduates. For the other specialties, we believe a lower overall HCD cost per
unit of productivity can be achieved by increasing course length. (See
pp. 45–49.)

The seven specialties specifically examined in this study offer
prototypes for our overall methodology. One essential element is
the derivation of productivity curves. The Air Force's Occupational

Table S.1
Analysis Summary

AFSC	Specialty Title	Write-In Comments	Add vs. Drop Comparison	Incremental Change Functions	Steady State Analysis
1A8X1	Airborne Cryptologic Linguist	Don't add	Drop	Decrease	
1N3XX	Cryptologic Linguist		Drop	Decrease	
2T3X1	Special Purpose Vehicle Maintenance			Increase	Increase
2A3X3A	F-15 Tactical Aircraft Maintenance			Increase	Increase
2A5X1E	B-1/B-2 Aerospace Maintenance	Don't drop	Add	Increase	Increase
3P0X1	Security Forces	Don't drop	Add	Increase	Increase
3E7X1	Fire Protection		Add	Increase Don't decrease	Increase

Measurement Squadron (OMS) measures certain aspects of productivity through its occupational measurement surveys. We recommend that AETC investigate the use of current OMS tools and potential enhancements to develop productivity curves and the functional relationship between incremental changes in IST content, course length, and graduation effectiveness. We also recommend that the AETC Studies and Analysis Squadron use our HCD cost model and methodology for future IST course length studies. Other specialties can be readily examined. Improvements in methodology might include greater fidelity in generating the productivity curves, determining the impact of curriculum changes, and estimating costs. (See pp. 30–35 and 57–64.)

Finally, this analysis approach offers great potential for trades. In three of the AFSCs (2T3X1, 3P0X1, and 3E7X1), a 0.5-percent change in total costs for increasing course length resulted in a tenfold (5.0-percent) change in total productivity. (See pp. 63 and 67.)

In summary, our analysis suggests the following:

- *A significant increase in productivity for small addition in IST course length* is likely to exist for five of the seven specialties we analyzed. (See pp. 63 and 67.)
- In some cases, *portions of the IST curriculum could be reduced with little impact on productivity.* We believe there is evidence from the qualitative responses for considering reductions in the 1A8X1 and 1N3XX specialties. (See pp. 47–49.)
- Although we believe these increases are plausible given the way we conducted the analysis, *it would be prudent to replicate the data with more specific and refined survey estimates.* (See pp. 70–71.)
- *The Air Force should also investigate other external measures* to validate the productivity functions we derived. (See pp. 23–31 and 71.)
- This analysis also demonstrates the *large role played by the cost of OJT and its importance in policymaking,* particularly for determining the course length of IST. (See pp. 62–68.)
- We recommend that *the specific results of this analysis should be briefed to the respective AFSC U&TWs* as they consider the suitability of the current IST curricula. (See pp. 57–68.)
- Finally, we recommend that *the AETC Studies and Analysis Squadron (SAS) adopt the models and methodology* developed in this study for future analyses involving the length of IST. (See pp. 57–68.)

Acknowledgments

We wish to thank Lt Gen (ret.) John D. Hopper, Jr., (then Vice Commander, AETC) for the inspiration for this study and his enthusiastic support. Mike Snedeker (AETC/AXP) did a lot of the legwork to launch the study and ensure that a useful set of specialties was evaluated. Col Rick Naylor (AETC/DOO) and his staff provided expertise and open doors for literature, subject matter expert (SME) access, and database needs.

We also are grateful to Lou Datko (AFPC/DPAFFA), who sent out email notices to our survey population. Tom Bogdon and Christopher Corey of the RAND Survey Research Group did an outstanding job in developing an online survey tool for the study.

We wish to thank the AETC Studies and Analysis Squadron (AETC SAS). Our special thanks to the Commander, Lt Col Lee Williams, and to Capt Paul Clemans, who helped us gain permission from ACC and AMC to survey their personnel.

Clifton (Corky) Scribner (360 TRS/TRR) deserves special recognition for his work in supporting our team when we visited Sheppard Air Force Base. We were assisted by numerous subject matter experts from a variety of locations within AETC. We want to thank CMSgt Tim Steffen (AF/XOOTA), SMSgt Scott Lawson (AFPC/DPAOM4), CMSgt Ernest D. Shishido (HQ AIA/DOXF), Lt Col James Sohan (312 TRS), CMSgt Terry Ford (312 TRS), Scott Hebert (312 TRS), MSgt Tony Lohrman (312 TRS), MSgt Ron Prettyman (312 TRS), Santos Zamarripa (343 TRS), MSgt Daniel Specker (345 TRS), MSgt Michael Phillips (37 LRS), MSgt Todd Huckaby (316 TRS), Maj

Steven Anderson (316 TRS), Maj Jonas Skinner (316 TRS), MSgt Paul McClelland (316 TRS), Lt Col Mack Breeland (360 TRS), and others who served as subject matter experts and/or helped with the dry run of our survey.

Finally, we wish to thank our reviewers, Col (ret.) Danilo Medigovich and Michael Polich for their suggested changes, corrections, and improvements. Their comments greatly improved the report.

Abbreviations

1A8X1	Airborne Cryptologic Linguist AFSC
1N3XX	Cryptologic Linguist AFSC
2A3X3A	F15 Maintenance AFSC
2A5X1E	B1/B2 Maintenance AFSC
2T3X1	Special Purpose Vehicle Maintenance AFSC
3E7X1	Fire Protection AFSC
3P0X1	Security Forces AFSC
ACOL	annualized cost of leaving
AETC	Air Education and Training Command
AETC/CV	Vice Commander, Air Education and Training Command
AETC/DOR	Air Education and Training Command, Directorate of Operations
AETC SAS	Air Education and Training Command Studies and Analysis Squadron
AF/DPL	Directorate Learning and Force Development, Headquarters Air Force
AFI	Air Force instruction
AFOMS	Air Force Occupational Measurement Squadron
AFSC	Air Force specialty code
APOM	Amended Program Objective Memorandum
ATI	automated training indicator

BMT	Basic Military Training
CFETP	Career Field Education and Training Plan
CFM	career field manager
E1 thru E9	enlisted grades (lowest to highest)
ETS	expiration of term of service
FEQ	Field Evaluation Questionnaire
FTD	Field Training Detachment
FY	fiscal year
GAS	Graduate Assessment Survey
HCD	human capital development
IST	initial skills training
IQR	inter-quartile range
MAJCOM	major command
O/I	organizational/intermediate
OMS	Organizational Management System
OJT	on-the-job training
ORI	Operational Readiness Inspection
OSD	Office of the Secretary of Defense
OSR	Occupational Survey Reports
PAF	Project Air Force
PME	professional military education
POI	plans of instruction
POM	Program Objective Memorandum
SME	subject matter expert
SAF/FMC	Deputy Assistant Secretary of the Air Force for Cost and Economics
TT	technical training
U&TW	utilization and training workshop
YOS	year of service

CHAPTER ONE

Introduction

This monograph examines the cost and readiness trade-offs between centralized initial skills training (IST) and decentralized (on-the-job) training (OJT) in order to assist training managers and other decision makers in determining the preferred length of initial skills training (basic military training [BMT] to 3-level award[1] for mission-ready airmen). Specifically, we examined seven specialties: Airborne Cryptologic Linguist (Air Force Specialty Code [AFSC] 1A8X1); Cryptologic Linguist (AFSC 1N3XX); Tactical Aircraft Maintenance, F-15 (AFSC 2A3X3A); Aerospace Maintenance, B-1/B-2 (AFSC 2A5X1E); Special Purpose Vehicle Maintenance (AFSC 2T3X1); Fire Protection (AFSC 3E7X1); and Security Forces (AFSC 3POX1).

To determine whether IST course lengths should be increased or decreased, we needed to assess how rapidly the current training program prepares airmen in the different specialties to gain the skills necessary to be fully mission capable. Rather than examining individual skills for each specialty, we created the more general concept of a *fully productive airman,* one who possessed the skills to be fully mission-effective. This led us to the following research questions:

1. At what level of productivity do airmen currently graduate from IST?
2. How does productivity increase with time in OJT?

---

[1] The 3-level award occurs at graduation from IST.

3. How do manning limitations and deployments affect the quality of OJT?
4. Given the current IST curriculum, what skill elements should be added or dropped, and what would be the impact on airman productivity at IST graduation and throughout their career?

To address these research questions, we fielded and analyzed an extensive survey of senior enlisted personnel in each of the specialties. We also created an analytic framework to examine the cost-benefit relationship of changing the length of IST on the total cost for human capital development (HCD). This analytic framework can be tailored for other Air Education and Training Command (AETC) courses, such as supplemental technical training, flying training, and educational courses.

Motivation and Background

AETC trains roughly 30,000 to 40,000 individuals a year in initial skills. The average length of training at an AETC base, including BMT, is 23 weeks, with an average cost of $20,000 per student. A rough estimate for the overall cost for initial skills training is in the neighborhood of $700 million. The cost of OJT can approach $40,000 per airman or $1.4 billion a year in OJT costs.[2] Beyond the costs, each type of training delivery has various advantages and disadvantages, which we discuss in more detail below. Given these parameters, the primary study question is, What is the optimum mix of IST and OJT that either maximizes unit readiness for a given cost or minimizes cost for a specified level of productivity?

Typically, training reviews conducted in the past have not fully evaluated how changes in IST affect the costs and other burdens of OJT. For example, in July 2003, the Air Force Council requested two training studies (later referred to as the AF Council–Directed Train-

---

[2] Oliver et al. (2002) estimated OJT trainee and trainer costs in the range of $31,000–$68,000 for aircraft maintenance AFSCs. The average maintenance AFSC OJT cost was $40,000.

ing Review). The first study looked at a one-day reduction per year for three years in each AFSC in order to achieve identified fiscal year 2005 (FY05) Amended Program Objective Memorandum (APOM)[3] manpower and dollar adjustments. The second study was a larger review of flying training, technical training, and education courses. The Air Force Directorate of Learning and Force Development (AF/DPL) led the FY05 APOM study and AETC led the FY06–FY07 Comprehensive Training Review. Neither of these efforts addressed the effect of adding or reducing specific pieces of course material or the overall impact on airman productivity and cost after graduates reach the field. Similarly, commonly available estimates of technical training costs do not include the full costs of OJT training—partial productivity of the trainee, impact on supervisors, the effect of supervisor-to-trainee ratio, and the disruption of training due to deployments.[4]

To permit a fuller consideration of overall human capital development costs, the Vice-Commander of Air Education and Training Command (AETC/CV) and RAND Project AIR FORCE (PAF) conceived another approach to evaluating training reductions. Building on PAF work that used productivity estimates to evaluate the costs and benefits of reenlistment bonuses, we identified a similar approach to finding the best mix of IST and OJT. A key part of this analysis was including OJT costs as part of the full HCD cost. Additionally, instead of looking at marginal cuts to course lengths, we examined eliminating or including whole skill areas, regardless of the length of time necessary for instruction. Our methodology provides a means to optimally balance the costs of IST with the costs of OJT. This approach allows

---

[3] APOM is an out-of-cycle Program Objective Memorandum (POM) amendment.

[4] Training disruptions due to deployments can occur in at least one of three ways. First, the trainer can be deployed without the trainee. This leaves the trainee at the home station with different and potentially less experienced trainers to continue OJT. Second, the trainee can be deployed without the trainer. OJT must then continue in the field with different trainers. Finally, trainee and trainer can both be deployed. For some deployments, the experience in the field will not align with OJT needs. In other cases, the deployment provides opportunities to exercise job skills that could not be performed at home station. Nonetheless, deployments do create disruptions in ongoing OJT.

us to calculate the course length that provides the lowest overall HCD cost for a specified level of productivity.

Centralized and Decentralized Training in the Air Force

Centralized and decentralized training are two different, general approaches for providing the skills airmen require in order to become fully mission ready. Centralized training is primarily provided by AETC as IST; decentralized training occurs as OJT within units under each major command (MAJCOM).

AETC IST has the advantage of providing dedicated training and a concentrated learning environment where skills can be taught in a highly standardized fashion. IST provides greater uniformity of instructors and training devices, but must pull manpower and equipment completely away from operations. In IST, no work is produced that adds directly to mission readiness. IST can utilize hands-on training but requires training devices (some of which are specialized trainers or actual equipment).

MAJCOM OJT training allows better workstation-specific skill development and potentially an earlier contribution of the trainee to actual operations and readiness. In OJT some productive work is accomplished, but training often occurs in increments over longer periods of time. OJT is a mixture of hands-on training and independent study. Furthermore, OJT allows trainees access to actual supervisors and unit-specific equipment, although this takes both trainees' and trainers' time away from normal duties.

For each AFSC, then, the primary question is which skill development tasks, and the associated required resources, should be assigned to AETC IST and which should be assigned to MAJCOM OJT.

Three components are needed for this decision: determining skill achievement, associating it with productivity in terms of mission readiness, and estimating training costs. First, skill achievement must be defined and measured. This is already partially captured in a variety of tests and surveys—the Graduate Assessment Survey (GAS), the Field Evaluation Questionnaire (FEQ), field interviews, and work done by the Air Force Occupational Measurement Squadron (AFOMS). For our purposes, we needed to derive learning curve distributions that relate training to skill achievement and effectiveness across the range of trainees. Unfortunately, the current set of tests and surveys measures achievement at a level of detail that is difficult to aggregate into a more general measure of job effectiveness.

Second, skill achievement must be related to readiness standards. Some methods exist, but consolidating, standardizing, and extending them is necessary. (As we have noted, we equate productivity achievement with the development and mastery of skills in a specialty necessary to be fully mission ready.)

Finally, an estimate is required of the resources (money, time, and people) necessary to teach skills in a centralized IST or a decentralized OJT manner. Some of the costs are straightforward. Some, like the capital costs for facilities or training devices, are more difficult to capture on a per-student basis. Furthermore, synergisms among some skills within a specialty may make it less costly to teach them together rather than separately.

Once productivity and the numerous types of costs are determined, they can be combined to determine both mission readiness and overall cost. Figure 1.1 shows a notional graph of productivity over time for an airman. The horizontal axis shows years of service for an individual. The vertical axis is the relative productivity of the individual. In this example, the individual completes IST in about five months. During IST, the trainee contributes nothing in terms of productive work; rather, all the airman's time is spent in skill development. Thus, all the pay, allowances, and other costs to support the airman, along with the costs of training, are attributed to human capital development (depicted in the figure as the area marked "Initial Skills Training").

Continuing with Figure 1.1, after graduation from IST (3-level award), the airman enters the force at approximately 20 percent of the productivity of a fully qualified journeyman and is assigned to an operational unit. Training does not stop at this point; it still occupies a significant portion of the airman's time. OJT allows an airman to spend some time performing productive work while learning valuable

Figure 1.1
Capturing the Full Cost of Human Capital Development

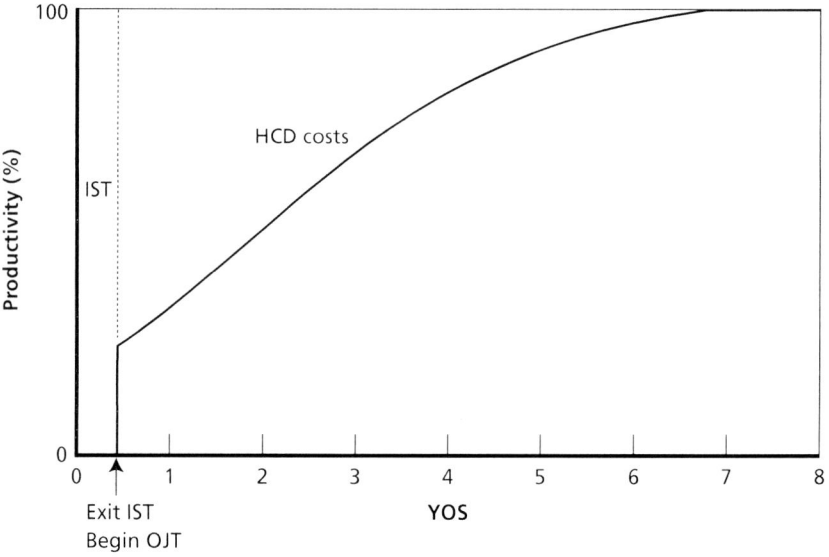

skills in an operational environment. Eventually, the individual builds up to 100-percent effectiveness—by year seven, in this example. Unlike with IST, the trainee is able to contribute to the effectiveness or readiness of the unit while undergoing OJT.

Too frequently, the costs associated with OJT are incompletely accounted for. The true costs of training are not limited to those associated with IST and advanced courses, but should include such OJT costs as

- the fraction of time the airmen is learning rather than contributing to the mission of his or her unit
- time when the airman is only partially effective because of underperformance or mistakes
- additional equipment or consumables needed for training or for repairing mistakes
- supervisor time needed for instruction or to correct mistakes.

Thus, our definition of full HCD costs includes not only the commonly reported cost of IST, but also OJT costs.

Considering full HCD costs allows us to optimally place the line between IST and OJT. Given alternative productivity curves, we can shift the line at which the individual exits IST and examine the impacts on both productivity and cost. When we do this, we observe a number of interesting changes.

In Figure 1.2, the course length for IST is shortened. The reduction in IST also lowers the instructor requirement and reduces accessions because fewer man-years are used for training. The shift in man-years to actual operational manpower slots reduces the yearly input requirement.

However, this change will increase OJT costs. By exiting IST early, the individual enters the force with a lower overall productivity.

Figure 1.2
The Effect of Shorter IST Course Length

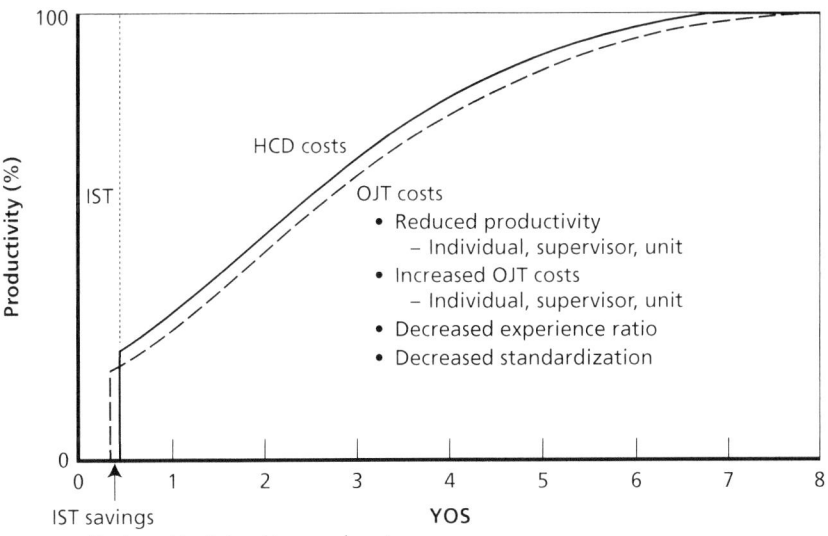

The lower productivity increases OJT costs. With more inexperienced personnel in the units, the supervisor-to-trainee ratio also decreases, thus taking more productive time away from the supervisor and reducing the effectiveness of OJT. Finally, as more training is transferred to the unit, there is an increased risk of less standardization in the training, because the multiple OJT locations will tend to train to more of their own individual methods and needs.

On balance, then, should IST be shortened? The complexity of the multiple interactions discussed above makes it difficult to determine if the costs and benefits favor less IST or more.

Similarly, we could increase the time in IST and would see an opposite set of effects. A full understanding of both IST and OJT costs and their interactions is required to determine the optimal IST length. The framework we propose captures and minimizes total HCD costs per unit of productivity. The key element is determining the productivity function and the impact of a change in IST on the productivity function. This challenging problem is the primary analytic task of this monograph.

An Approach for Determining the Length of Centralized Training

Our methodological approach consists of four steps:

1. Collect specialty-specific training data—retention, pipeline courses, plans of instruction (POIs), career field education and training plans (CFETPs), occupational survey data, additional courses, typical OJT tasks, and associated costs.
2. Develop skill-based learning curves for each specialty. To do this, we conducted surveys and interviews with supervisors and subject matter experts (SMEs).
3. Estimate resource requirements in terms of dollar cost, manpower, devices, facilities, and time.

4. Develop productivity-versus-cost curves for each specialty and determine the preferred allocation of skills to centralized and decentralized training.

AETC asked us to examine seven Air Force specialties representing a broad spectrum of career fields. We used these specialties to test our methodological framework. The first two, Airborne Cryptologic Linguist (AFSC 1A8X1) and Cryptologic Linguist (AFSC 1N3XX), are among the longer pipeline courses (over a year). The next three are maintenance skills: Tactical Aircraft Maintenance, F-15 (AFSC 2A3X3A), Aerospace Maintenance, B-1/B-2 (AFSC 2A5X1E), and Special Purpose Vehicle Maintenance (AFSC 2T3X1). The Fire Protection specialty (AFSC 3E7X1) requires civilian levels of certification. Last, the Security Forces specialty (AFSC 3P0X1) is currently experiencing many new challenges brought on by a rapidly changing warfighting environment.

Our methodology is easily adapted to other AFSCs. Examining the seven initial AFSCs was helpful in testing our approach. Later in the monograph, we discuss options for improving the methodology for determining the most cost-effective course content and length.

Data Sources

A significant task in this study was identifying and obtaining requisite data for our analysis. Table 1.1 provides a summary of our data sources and needs. A variety of course-related data was obtained from AETC in order for us to better understand and represent IST and follow-on training in specific specialty areas. The CEFTPs allowed us to examine individual course skills and goals in some detail. AETC also provided the fundamental data on course lengths and sequences. We used washout and washback rates to help estimate the cost of training.

Table 1.1
Data Sources

Data Source	Data
Air Education and Training Command	CEFTPs Course sequences Course data, lengths Washout, washback rates
Occupational Measurement Squadron	OSR[a] data (tasks, difficulty, training emphasis)
Air Force Personnel Center	Proportion of population at end of ETS[b] Proportion of population not at end of ETS Reenlistment rates (Pre- and post-9/11) Continuation rates Crossflow by YOS[c] and grade Promotion rates by YOS and grade Course information (% attending, timing) Populations by YOS and grade Manpower authorizations
Air Expeditionary Force Center	Deployment cycle data
Air Force Instruction 65-503	Detailed wage costs Inflation rates IST course costs

[a]OSR = Occupational Survey Report.
[b]ETS = expiration of term of service.
[c]YOS = years of service.

We initially analyzed Occupational Survey Report (OSR) data to build productivity curves. We found, however, that the OSR data were insufficient for our needs and determined that a survey was the best method to develop the productivity curves and the effect of changes to IST. We did use the OSR data to get an understanding of the skill areas in each specialty that were "on the edge" of inclusion or exclusion from IST.

As part of our methodology, we simulated the composition of the entire enlisted force out into the future. Therefore, we required not only training-oriented data, but also a variety of personnel data to make these projections. The Air Force Personnel Center supplied most of these data. The model required the percentage of personnel who are at the end of their term of service, their reenlistment rate, and the continuation rate for those not at the end of their term of service.

We also used cross-flow rates (movement between AFSCs) by years of service (YOS) and grade. The model used actual promotion rates and the percentage of time in training after initial skills training. We also used manpower data to examine the discrepancies between manning and authorizations.

Additionally, data from the Air Expeditionary Force Center allowed us to represent deployment cycles and their impact on OJT.

Cost represents another significant dimension of our study. Variable IST course costs, composite wage costs, and inflation rate data were obtained from Air Force Instruction 65-503 (Secretary of the Air Force, 1994b).

Missing from this list of data is the information necessary to build the productivity curves described above. The survey instrument used to gather this information is a key part of this study and is discussed in detail in later chapters.

Structure of the Monograph

Chapter Two is a review of the literature pertaining to the relative effectiveness of technical and on-the-job training, estimating worker productivity, and deriving costs. This literature includes past studies, current training and career field guidelines, and available data sources. Chapter Three describes our data collection instrument and provides the results of the survey. It includes the development of our productivity curves and functions representing productivity change due to increases and decreases in technical training course length. Chapter Four discusses the elements of cost associated with both IST and OJT. In Chapter Five, we combine the productivity and course length curves with cost to determine the length of initial skills training that provides the greatest benefit per unit of cost. Each specialty is considered separately to capture its unique aspects. Finally, in Chapter Six we discuss our recommendations for the selected specialties and make a variety of suggestions for improving information collection and the analytic methodology.

CHAPTER TWO

Review of the Literature

In this chapter, we review previous attempts to measure productivity to determine the relative capabilities of new IST graduates, calculate the cost of OJT, and estimate the effect of shortening or lengthening IST. Additionally, we describe the current Air Force training development process, as well as some of the instruments used to assess the quality of the training program. Finally, we briefly discuss our study in the context of these other studies.

Measuring Productivity

Defining and measuring productivity is a significant challenge in the military environment. Our methodology was based on a survey instrument designed to measure trainee productivity relative to that of a carefully defined fully productive airman. Our definition of productivity and the questions for our survey instrument build on previous research.

Haggstrom, Chow, and Gay (1984) developed a widely used survey instrument entitled the Enlisted Utilization Survey, which was a large-scale survey of trainee supervisors that attempted to measure the productivity of enlistees over time. Survey respondents were asked to rate the net productivity of their specific trainees relative to the typical individual with four years of experience in the same specialty at four distinct time points. Respondents were also asked to rate the net productivity of an "average trainee" at the same points. One particularly interesting finding from this study was that supervisors tended to rate their own trainees higher than the average trainee, indicating that using the "average trainee" values from a similar estimate might produce a downwardly biased estimate of actual OJT training costs.

Horowitz and Angier (1985) used an econometric time series regression technique to measure the relationship between experience (measured by rank, years of service, prior sea experience, and quantity of training) and productivity (measured by the amount of mission-degrading downtime suffered by the maintained equipment) for individuals in six Navy maintenance career fields. Although their techniques were not designed to demonstrate causal relationships between any of the experience variables and productivity, their analysis did indicate significant correlation between quantity of formal training and productivity in five of the six career fields in the study. All six career fields indicated a significant relationship between rank and productivity. These findings were presented in the context of a broader report that emphasized the composite benefits of the Navy's extensive OJT training program.

Oliver et al. (2002) developed productivity curves for aircraft maintenance specialties by surveying a small number (20–30) of SMEs in each maintenance specialty. They defined a fully mission-effective person as 100-percent productive when the individual was considered a "go-to" person by the leadership. Oliver et al. used the productivity curves and an annualized cost of leaving (ACOL) model to evaluate various incentive programs. They took the cost of the different incentive programs, the resulting changes in attrition, and the overall productivity to show the optimal cost-productivity relationship. The use of productivity curves and the ACOL model allowed them to show the payoff from implementing incentive programs for maintainers. We, too, used this type of survey to determine productivity curves for each specialty.

Comparing Technical Training Graduates and Directed-Duty Assignees

The vast majority of studies examining the relative efficiency of IST and OJT have looked at differences between *directed-duty* and *regular* enlistees. In some situations, the military services use a directed-duty assignment program whereby selected enlistees bypass IST and receive all their upgrade training on the job. The research question for these studies tended to focus on the differences between individuals who had received OJT exclusively and individuals who attended an initial skills course prior to completing upgrade training on the job. Results of those studies were mixed.

Weiher and Horowitz (1971) compared the relative costs of formal and on-the-job training for Navy enlisted occupations. They gathered data for their study via a questionnaire in which respondents were asked to estimate how long it took the average trainee to reach "third-class" level and how much time senior personnel spent on OJT instruction. This study also attempted to develop productivity profiles of A-School (the Navy equivalent of IST) graduates and directed-duty enlistees by having respondents draw a separate curve for each, showing the proficiency over time of a typical trainee relative to that of a man "professionally qualified to take the third-class exam." These data were then aggregated to derive a curve for the typical man in each training mode. The authors used the survey data to generate estimates for school costs, value of foregone productivity, and supervisor costs for each of the training nodes. Assuming that the estimates from the survey data were correct, the results indicate that when all the relevant costs are considered, 37 out of 39 career fields are more efficiently trained with A-School. The authors emphasize, however, the importance of accurate estimates of "supervisor costs" in generating the apparent efficiency.

Smith (1986) conducted an observational study of first-term airmen to determine if there was a significant difference in productivity between technical school graduates and directed-duty airmen in each of their first four years of enlistment. In this study, productivity was defined as the "proportion of productive work-time" based on a week-long series of personal observations. The study found no significant difference between the productivity of the two groups during years two, three, and four, but found that "first-year" technical training graduates spent a significantly greater proportion of time doing productive work. Smith used these proportions, estimates of lost trainer productivity due to OJT (from SME interviews), and estimates of the cost of technical training (published Air Force data) to compare the total costs of each method. His final results indicated that the total cost to the Air Force for directed-duty airmen was $18,000–$43,000 more expensive (in 1986 dollars) than the total cost for producing technical training graduates.

Quester and Marcus (1986) used Navy data from an Enlisted Utilization Survey to compare the productivity of A-School graduates versus directed-duty assignees in seven Navy specialties in order to compare the cost-effectiveness of the two types of training. Productivity, in this study, was defined as a "net productivity"—the contribution of the trainee minus the loss in production of the experienced personnel who must train and supervise him. Quester and Marcus found that the productivity of A-School graduates relative to directed-duty assignees, over the course of their entire first-term enlistment, ranged from 1.16 to 1.41, and that the total cost per unit of effectiveness was lower for A-School in four of the seven fields, suggesting that A-School is more cost-effective.

Calculating Costs for OJT

Three studies provide costing methodologies for OJT. In the first, Carpenter-Huffman (1980) described three types of OJT activities and the importance of quantifying each when calculating the full cost of OJT. Type I activities are those whose only product is training, Type II activities are those that trainees can do without any additional training, and Type III activities are those that produce work and training simultaneously. She suggests a methodology that calculates the differences in the costs of resources required by OJT performing units and non-OJT performing units (that produce otherwise equivalent work), to determine the cost of providing the OJT training. Carpenter-

Huffman only proposed this methodology—no known study was ever completed using it.

In the second, Fleming et al. (1987) developed a methodology to quantify the cost of OJT training in their effort to calculate the replacement cost of fully trained Air Force personnel. They calculated the costs of unproductive trainee time during OJT training periods and the costs of unproductive trainer/supervisor time during those same periods. Using data provided by the Air Force Operational Measurement Squadron survey reports on the percentage of time spent (by grade) on OJT-related tasks, the average wages for each grade, and the assumption that the rate of productivity acquisition can be estimated in a straight-line function, the authors were able to generate a cost value for the unproductive OJT time. Using their methodology, the cost of the unproductive OJT time exceeded the published cost of technical training in 23 out of 37 specialties. Although this does not directly determine which training method is more effective, it is a strong indication that the cost of unproductive time due to OJT cannot be ignored.

Gay (1974) developed an OJT costing methodology based on an application of human capital development theory. In his methodology, the military's investment in OJT is measured as the "present value of the sum of positive differences between an individual's military pay and productivity over time." In his study, pay was measured as the expected value of military pay and allowances in the particular military specialty by length of service. Military productivity was measured by supervisors' estimates of the time required for individual trainees to reach readily identifiable milestones in their OJT performance. His fundamental conclusion from an aircraft maintenance specialist pilot test was, "A substantial portion of training costs is in the form of OJT costs which, although quite real, are not at present well identified." The estimated cost of technical training was $3,200 per trainee, and the estimated cost of OJT was $6,600 (both in 1974 dollars). Gay and Albrecht (1979) extended this methodology to an additional 50 specialties across the military services; although they focus on the validity of the time-based productivity curves they developed with their survey as opposed to any effort to determine the economically efficient amount of technical training.

With the current system that requires OJT for all airmen, regardless of specialty, during their first duty assignment, opportunities to implement a methodology similar to the one suggested by Carpenter-Huffman seem unlikely. There may be units that do not perform any OJT, but the data points would not be sufficient to do a widely generalizable study. The utility of Fleming's methodology, on the other hand, has been demonstrated using data regularly collected by the Air Force. Further development of this model with less restrictive simplifying assumptions, while beyond the scope of our current research, could be worthwhile.

Additional Options

Adjusting the IST/OJT boundary is not the only available approach for finding a more cost-efficient training pipeline. Unpublished 1994 RAND Corporation work by John Winkler provides an excellent discussion of potential training consolidation options, Hanser, Davidson, and Stasz (1991) give an in-depth analysis of civilian-provided training as an alternative to the current technical training (TT)/OJT system, and Kavanagh (2005) has published a literature review of the relationship of recruiting, advanced training, and retention to productivity. Our research did not investigate these policy levers. A comprehensive training pipeline analysis would certainly need to consider them.

Current Program and Available Data

Air Force Instruction (AFI) 36-2201 (Secretary of the Air Force, 2004) describes the current Air Force training development process, including the responsibilities of the career field manager (CFM), preparing and conducting a utilization and training workshop (U&TW), and constructing a career field education and training plan. The CFETP identifies training requirements, support resources, and core task requirements for a career field. One CFM is appointed in each career field to, among other responsibilities, ensure development, implementation, and maintenance of the training plans for his or her assigned specialties. The U&TW, which is "to be used as forum to determine education and training requirements, by bringing together the expertise to establish the most effective mix of formal and on-the-job training requirements for each AFS skill level," is one of the primary tools the CFM uses to evaluate and modify a career field CFETP.

In addition to their career field expertise, participants in a U&TW utilize a variety of data to make decisions about career field training, such as the GAS, the FEQ, and the OSR from AFOMS.

The GAS is used to gather customer feedback on active duty Air Force, Air Force Reserve Command, and Air National Guard graduates of officer and enlisted initial skill Type 3, 4, and 5 courses (Hostage, 2005, Para 5.5). GAS data are gathered from the supervisors of technical training graduates after they have been at their first duty station a minimum of 90 days, requesting feedback on the attitude and preparedness of each graduate. The results of these surveys are reported at the course, specialty, base, and MAJCOM level and are intended to provide a picture of the overall quality of the IST program.

The FEQ is sent to supervisors requesting data/feedback concerning recent graduate qualifications in approved training standards. The data are then used by the schoolhouses to update and improve the course curriculum.

The OSRs provide a wealth of information about each specialty, including detailed information about the difficulty and frequency of performance of each task within a CFETP, job satisfaction, and a predictive retention analysis. Of particular interest for our study were the detailed task analyses. These analyses are derived from three separate survey instruments—a general questionnaire given to a sample of all members of a specialty requesting information about the frequency with which tasks are performed and two more specific instruments given to senior members of the specialty requesting information about the importance and difficulty of each of those tasks. AFOMS reports information directly from the survey responses (the percentage of members performing each task, a list of all tasks in order of importance and in order of difficulty, and so on) and also calculates an automated training indicator (ATI) for each task. The ATI is derived from an extensive logic matrix, including criteria such as the percentage of first-term airmen performing a task, the average training emphasis rating, the average task difficulty rating, and whether or not the task is a critical safety item. Each ATI ranking corresponds to a suggestion for how extensively to train the task, and for which tasks OJT is more appropriate than centralized training.

What This Monograph Adds

Previous studies examining productivity, the relative capabilities of new IST graduates, the cost of OJT, and the effect of shortening or lengthening IST have raised several key issues. First, it is difficult to define and measure "productivity" and particularly difficult to create a universally accepted definition within a survey instrument. Our study built upon previous efforts to create a definition of 100-percent productivity that includes more intangible measures of productivity than the simple skill level certification found in earlier research.

Second, there is no opportunity within the current IST/OJT training system to use a quasi-experimental design to examine differences between directed duty and technical training graduates, as was done in previous research. Today, such comparisons would require implementing a complex experimental program within the existing training pipeline. Instead, we provide a methodology to measure the effects of incremental changes to the IST/OJT pipeline without having to perform costly experiments. We lay the groundwork for further investigations of this question without sacrificing any of the standardization that a single pipeline option provides.

Finally, the most significant limitation of previous studies has been the failure to include all relevant costs within the cost-benefit calculation. Although each of the individual studies discussed above captures some component of the total costs, no study we examined simultaneously considers the tangible resource costs and the long-term costs associated with changes in the initial productivity of technical training graduates. Our study uses a deterministic, steady state model to compare the costs and mission-effectiveness benefits of various IST/OJT combinations. One of the most important contributions of this research is that the cost-effectiveness comparison considers the full range of costs within the calculation.

CHAPTER THREE

A Survey to Determine Productivity and Effectiveness

As noted in the introduction, the key missing information needed to analyze the trade-off between IST and OJT is a way to capture the productivity of airmen at the time of IST graduation and throughout their career. One way to approximate productivity is through expert opinion. This chapter describes the mechanism we developed for capturing the salient views of subject-matter experts.

Survey Contents

Our primary tool for gathering expert opinion was a twenty-question survey (see Appendix B) that spanned a number of topics related to job productivity. The survey was designed to be completed in less than half an hour. Table 3.1 outlines the question areas that we investigated.

The notion of a fully effective worker establishes a key benchmark for the entire survey and for our analysis. Essentially, we are defining an individual a supervisor would "go to" for most important unit-related tasks. The fully effective worker represents 100-percent effectiveness on the productivity curves we created. Using SME recommendations and previous research (Oliver et al., 2002), we developed the definition shown in Figure 3.1. With this definition in hand, we asked supervisors to rate how effective individuals were at various points of time and at key phases. Specifically, we asked for the average rank and YOS at which a person reaches fully mission-effective status. We also asked

Table 3.1
Overview of Survey Questions

Question Area	Number of Survey Questions
Background	3
Defining a fully mission-effective worker	2
Progress to fully mission effective	3
Leadership/management effectiveness	2
OJT requirements of trainers	2
Trainer productivity and quality due to training workload and manning	2
Deployment effectiveness	2
IST course content alterations	2

for the earliest and latest point observed. The intent was not only to capture the mean but also to understand the variance among SMEs' opinions of airmen they have observed. In other survey items, we asked supervisors to define progress to a fully mission-effective status at specific year points (1, 3, 5, 8, and 12) and specific events (3-level,

Figure 3.1
Fully Mission-Effective Survey Question

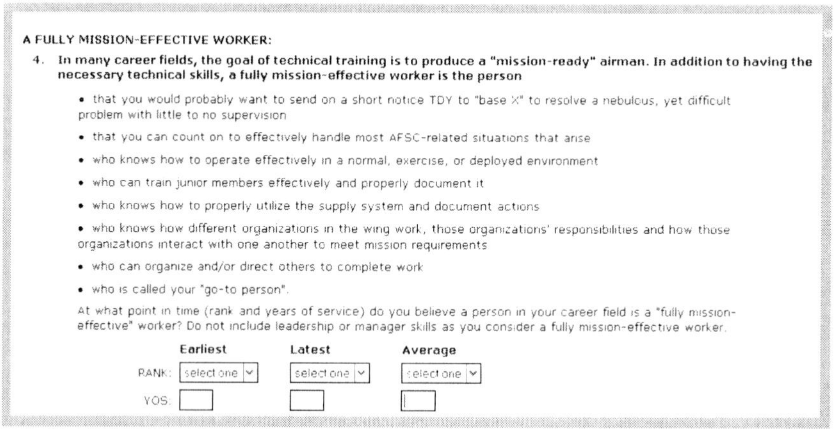
A FULLY MISSION-EFFECTIVE WORKER:

4. **In many career fields, the goal of technical training is to produce a "mission-ready" airman. In addition to having the necessary technical skills, a fully mission-effective worker is the person**

- that you would probably want to send on a short notice TDY to "base X" to resolve a nebulous, yet difficult problem with little to no supervision
- that you can count on to effectively handle most AFSC-related situations that arise
- who knows how to operate effectively in a normal, exercise, or deployed environment
- who can train junior members effectively and properly document it
- who knows how to properly utilize the supply system and document actions
- who knows how different organizations in the wing work, those organizations' responsibilities and how those organizations interact with one another to meet mission requirements
- who can organize and/or direct others to complete work
- who is called your "go-to person".

At what point in time (rank and years of service) do you believe a person in your career field is a "fully mission-effective" worker? Do not include leadership or manager skills as you consider a fully mission-effective worker.

	Earliest	Latest	Average
RANK:	select one	select one	select one
YOS:			

RAND *MG555-3.1*

5-level, and 7-level awards).[1] An alternative measure of fully mission-effective status—the rank and YOS at which an individual is ready to deploy—was also explored in the survey.

We recognize that there is a difference between a fully effective worker and a fully effective leader/manager. Although we surveyed the force for data on when a person is a fully mission-effective leader or manager, we did not use that information in our productivity calculations. Most of the productive work of a unit is done at the lower grades. Leadership and management are, in a sense, force multipliers. If we had included leadership and management, productivity would have dipped for the individual as he learns these new skills. Eventually, productivity would rise back as the individual became a fully mission-effective leader/manager. We did not want changes in productivity due to grade structure (at the higher grades) to obscure the issues we were examining between IST and OJT at the lower grades.

Referring again to Table 3.1, the second part of the survey focused on trainer-related effects. These included measurement of the consumption of trainer time at different points in the airman's development along with the actual and ideal rank and YOS of trainers, the effect of trainer productivity as the trainee-to-trainer ratio increased, and the impact of 5- and 7-level manning levels on OJT productivity and quality.

Finally, we asked for specific changes to initial skills training. We asked the respondents what they would add and what they would remove from IST. For each item, we asked how it would affect the length of IST and effectiveness of graduates.

Selection of Survey Participants and Sample Size

Table 3.2 provides information on the AFSCs, sample sizes, and number of responses. We sampled E-6s and above in each of the specialties. Because of the limited number of airmen in the cryptologic

---

[1] In the most general sense, levels are awarded for completion of training and experience acquired.

Table 3.2
Sample Size Populations and Responses

AFSC	Specialty Title	Number Sampled	Number of Responses	Response Rate (%)	Usable Responses
1A8X1	Airborne Cryptologic Linguist	130	110	85	105
1N3XX	Cryptologic Linguist	400	69	17	60
2T3X1	Special Purpose Vehicle Maintenance	400	137	34	129
2A3X3A	F-15 Tactical Aircraft Maintenance	400	135	34	125
2A5X1E	B-1/B-2 Aerospace Maintenance	400	157	39	153
3P0X1	Security Forces	400	110	28	100
3E7X1	Fire Protection	400	146	37	123

specialties, we also surveyed E-5s for AFSC 1N3XX and E-5s and E-4s for AFSC 1A8X1.

The survey participants were selected from four MAJCOMS: Air Combat Command, Air Mobility Command, Air Force Space Command, and Air Education and Training Command. Discussions with SMEs indicated that the selected MAJCOMs would provide an adequate sample of experiences for these specialties. We randomly selected survey participants for every AFSC except 1A8X1. Because of the small size of the 1A8X1 specialty, we selected the entire population.

Overall, the response rate averaged 34 percent.[2] However, for AFSC 1A8X1, we achieve a very high response rate (85 percent) because senior noncommissioned officers sent emails to the field requesting participation. The low response rate for AFSC 1N3X1 may be attributed to workload and the possibility that individuals in this specialty do not check their unclassified email accounts (if they have one). After cleaning the data (eliminating bad responses and respondents who had changed AFSCs), we obtained the usable responses shown in

---

[2]　While the response rate is normal for Air Force survey response, it is still low enough to be concerned with the potential for selection bias. One solution would be to institutionalize the survey as part of another accepted Air Force process, such as the OSR process.

Table 3.2. Appendix D includes a more detailed discussion of the data cleaning efforts.

Quantifying Time to Become Fully Mission Effective

Figure 3.2 presents the mean time specified by survey respondents to become the "go-to" person for each of the seven specialties. As expected, the two crew chief positions (2A3X3A and 2A5X1E) take the least amount of time. Also as expected, the cryptologist specialties (1A8X1 and 1N3XX) are among the longest. At first glance, the mean time for the fire protection specialty (3E7X1) seems too high. However, this specialty uses the same certification criteria required of civilian firefighters. In essence, the U.S. national firefighting requirements have defined a fully mission-effective person at a very high level, and those same criteria are used in defining fully experienced Air Force firefighters as well.

The Security Forces specialty (3P0X1) also seems high until one considers the changing work environment. Airmen in this specialty are expected to perform law enforcement, nuclear protection, base defense, and, more recently, rear area land maneuver operations. Consequently, the specialty has become more difficult, taking a longer time to reach fully mission-effective status.

Figure 3.2
Mean Time to Become Fully Mission Effective

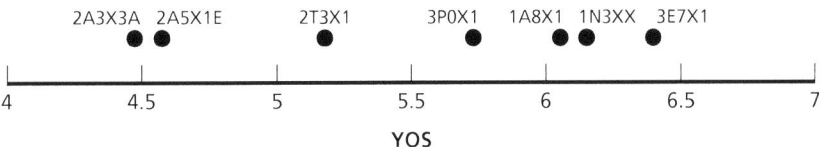

NOTE: 1A8X1 = Airborne Cryptologic Linguist; 1N3XX = Cryptologic Linguist; 2A3X3A = F-15 Maintenance; 2A5X1E = B-1/B-2 Maintenance; 2T3X1 = Special Purpose Vehicle Maintenance; 3E7X1 = Fire Protection; 3POX1 = Security Forces.
RAND *MG555-3.2*

Variance in Responses

Based on the respondents' definition of fully mission effective, the survey asked the participants to rate the productivity of the average airman on a scale of 0- to 100-percent of fully mission effective at 1, 3, 5, 8, and 12 years of experience.[3] The survey also asked the respondent to estimate productivity at 3-level, 5-level, and 7-level awards.[4] Figure 3.3 shows the high variation in the responses to the effectiveness of the average airman. The bars represent the middle 50 percent of the responses. For example, 50 percent of the respondents indicated that an airman at 3-level (IST) graduation is 10- to 30-percent effective. The lines extending from the boxes represent the middle 90-percent of the responses. Thus, 90 percent of the respondents believe that airman productivity at 3-level graduation is 10- to 40-percent effective. The square dot indicates the median response and the dash indicates the mean. So, as a whole, these SMEs believe that the median airman is 20-percent effective and the average airman is 15-percent effective at 3-level graduation.

The plot on the top shows the variability in response to the question of fully mission effective; the plots on the bottom include the time to 5- and 7-level. In this case, the middle 50 percent of the SMEs believe that the average airman takes four to seven years to become fully effective. Similarly, the middle 50-percent of the respondents believe that the average airman will receive the 5-level between three and four years and the 7-level between seven and eight years.

We were originally surprised by the large variance for effectiveness and productivity, particularly the long right tail for fully mission effective. We believe that two things contribute to this spread. First, there are no standards for how effective a person should be at any given time. Thus, it is unlikely that SMEs would have common reference points. Second, each respondent has a varied experience. This

---

[3] We also asked respondents to rate the productivity of the most effective airman and the least effective airman at each of these points. These extremes helped us to better understand the skill-level variations for each AFSC.

[4] The levels indicate the following: 3-level = apprentice, 5-level = journeyman, 7-level = craftsman, 9-level = superintendent.

Figure 3.3
Variability of Effectiveness Responses, AFSC 2T3X1

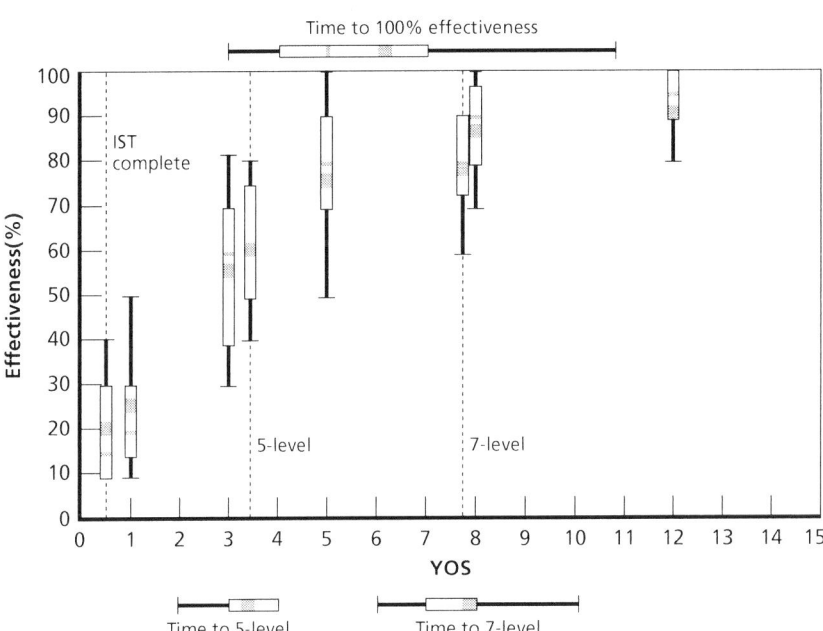

accounts for a natural variation that occurs because of the specific job requirements of a particular location. To measure the consistency of the data and the variability within our sample, we examined the standard error of the mean and the inter-quartile range (IQR). In Table 3.3, these statistical measures are presented for various years of experience in AFSC 2T3XX, including the average YOS at which the 3-, 5-, and 7-level awards are made.[5] For example, at 1 YOS, the average airman is, on average, 25.9-percent fully effective, and the IQR shows that the middle 50 percent of the respondents feel that the average airman is

---

[5]   The 3-level award happens at a known time based on the length of IST. The 5- and 7- level awards are based on individual accomplishment and can occur at a variety of times. Our survey also queried SMEs to determine, on average, at what YOS these awards occur. These averages are used in this table.

Table 3.3
Statistical Measures of Productivity Values, AFSC 2T3X1

YOS	Mean	Std. Dev.	Std. Error	IQR
0.4 (average at 3-level award)	20.9	14.0	1.3	20
1	25.9	15.0	1.4	15
2.4 (average at 5-level award)	61.1	15.6	1.5	25
3	56.5	20.4	1.9	30
5	76.0	18.0	1.7	20
6.8 (average at 7-level award)	79.0	12.0	1.1	15
8	87.0	13.7	1.3	17
12	92.2	10.1	1.0	10

15- to 30-percent fully mission effective. Table E.1 in Appendix E contains data for all seven specialties.

The standard errors are relatively small, indicating that the means fairly accurately represent the average effectiveness in this AFSC for each year of service or skill-level award point. Similar results are seen for all the AFSCs (see Table E.1). The IQRs are rather large, which shows the variability in the minds of a relatively small number of respondents. Again, we believe these results occur because of both the lack of definitive standards and the natural variation of experience among the respondents. It is also possible that some respondents did not understand the question, but few respondents indicated any difficulty in the remarks section of the survey.

Deriving Effectiveness Curves

When fitting curves to observations based on percentages that are not normally distributed and contain substantial number of observations at 0 or 100 percent, it is common practice to use a normalizing transformation of the data. To reduce the variability and better fit a curve to the data, we used an arcsine-square root function for this transformation (a full discussion is found in Appendix F). After the transformation, we employed a mixed effects model to fit a logistic growth model to the transformed data.[6] In order to bring the relationship back to

---

[6] A logistic growth curve is used to model functions that increase gradually at first, more rapidly in the middle growth perio d, and slowly at the end, leveling off at a maximum value after some period of time.

the original productivity scale, a reverse transformation was necessary. The last three columns of Table E.2 give the mean and the 95-percent confident interval after applying the reverse transformation.

Table E.3 provides the detailed parameter values (B_1, B_2, B_3, σ_u^2, and σ_ε^2) from the logistic growth model for each specialty. From these values, the productivity curves were generated. For each of these parameters, the table provides the mean value, confidence interval, and measures of statistical significance in the model. Of particular note is the small p-value for all the parameters indicating that in every case we have statistically significant values. This indicates that each of the parameters in the estimated equation is not zero.

Figure 3.4 is the result of fitting a curve to the predicted data for the 2T3X1 specialty. The upper and lower lines represent a 95-percent confidence interval about the predicted means. The tight confidence intervals for the predicted values provide assurance that, despite possible random processes, we have an accurate rendering of the true points on the curve.

Figure 3.4
Fit of Survey Productivity Data, AFSC 2T3X1

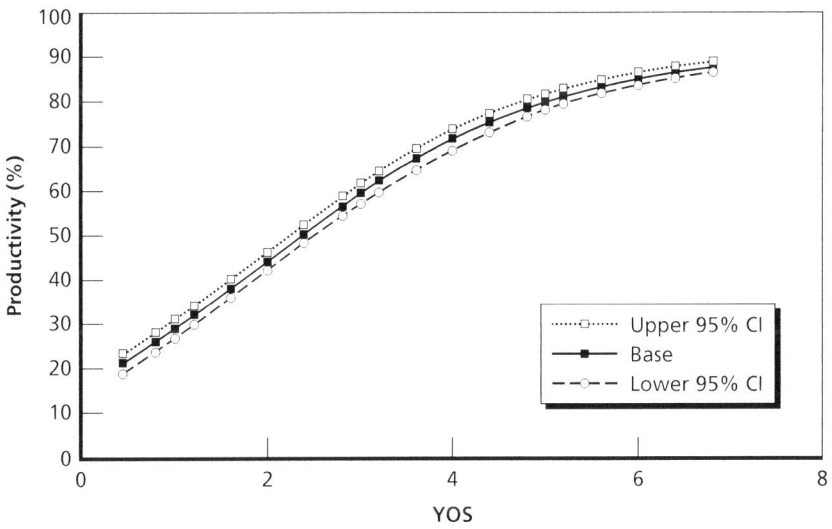

Scaling the Effectiveness Curves

As noted above, questions relating effectiveness and years of experience were asked in two different ways. First, we started with a fixed level of effectiveness (100-percent) and asked respondents to determine how long it takes to achieve that level. Second, we asked what level of effectiveness is achieved for a fixed number of years of experience. The responses to these two kinds of questions lead to potentially inconsistent results. For example, as shown in Figure 3.3, the median response regarding the time required to reach 100-percent effectiveness is about six years. However, when asked what the level of effectiveness is at 12 years, the median response is only about 90-percent.

Because this inconsistency occurred across all the specialties, we believe that it reflects respondents' willingness to attach a number of years of experience to 100-percent productivity but unwillingness to attach 100-percent effectiveness to any specific years of experience. From this we infer that when respondents focused on years of experience, they could always think of something more that airmen could learn that would increase their effectiveness. We believe that these additional skills are mostly related to location-unique needs and leadership or management tasks. Although important, the latter skills are not within our definition of the fundamental skills necessary for a 100-percent effective worker.[7] We therefore scaled the productivity curves to the years of experience required for 100-percent effectiveness in each of the respective specialties.

In Figure 3.5, we have scaled the productivity curves shown in Figure 3.4 such that an individual becomes fully mission effective (100-percent productivity) at the "go-to" point (Figure 3.2), the average amount of time to reach 100-percent effectiveness. For AFSC 2T3X1, the average time to 100-percent productivity, as given by the respondents, is about five years. This aligns with the approximately 80-percent productivity level specified by respondents when they were asked the question in terms of years of service. Thus, we scaled the produc-

---

[7] We ran excursions where the productivity curves were scaled to the 7-level point instead of the "go-to" point. The results, shown in Chapter Five, hold independent of the point to which the curve is scaled.

Figure 3.5
Scaling the Productivity Curves to the "Go-to" Point, AFSC 2T3X1

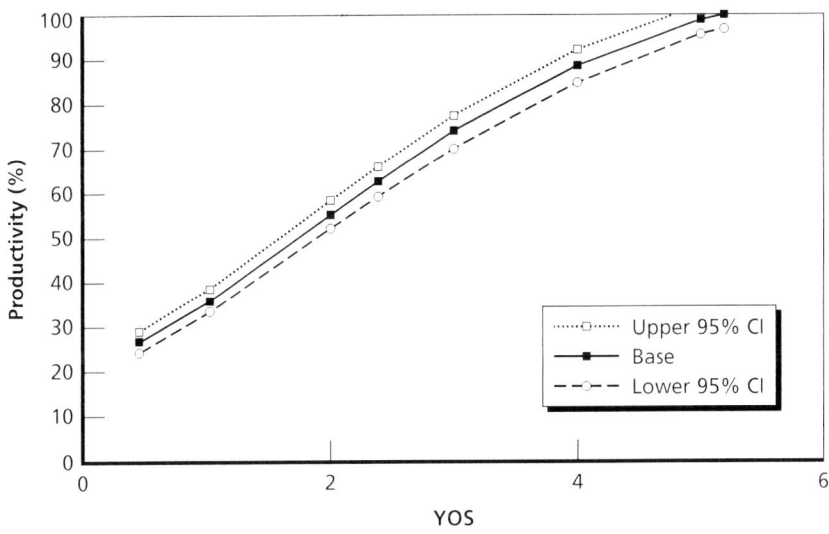

RAND *MG555-3.5*

tivity curve to reach 100-percent effectiveness around the fifth year.[8] The same transformation was performed for each of the specialties and used in the subsequent analysis. In the other specialties, the "go-to" point range was 75- to 84-percent productivity, with the average around 80 percent.

In Figure 3.6, we show all seven of the fitted productivity curves. These curves begin at the time of graduation from IST and extend upward until 100-percent productivity is achieved. The cryptologic AFSCs, 1A8X1 and 1N3XX, have significantly longer IST periods than the other five specialties, and so their curves begin farther to the right and at a lower level of productivity despite the longer IST length. This is primarily due to the difficulty of the foreign language skill they must learn. Also of note is that extending these curves back toward the origin gives a rough idea of the aggregate rate of learning during IST. Since the slope of this extension and the slope of the OJT effective-

---

[8]   We scaled 80-percent to 100-percent by multiplying 80-percent by 100/80 (or 1.2). We multiplied all the original values by the same factor.

Figure 3.6
Comparison of Fitted Productivity Curves

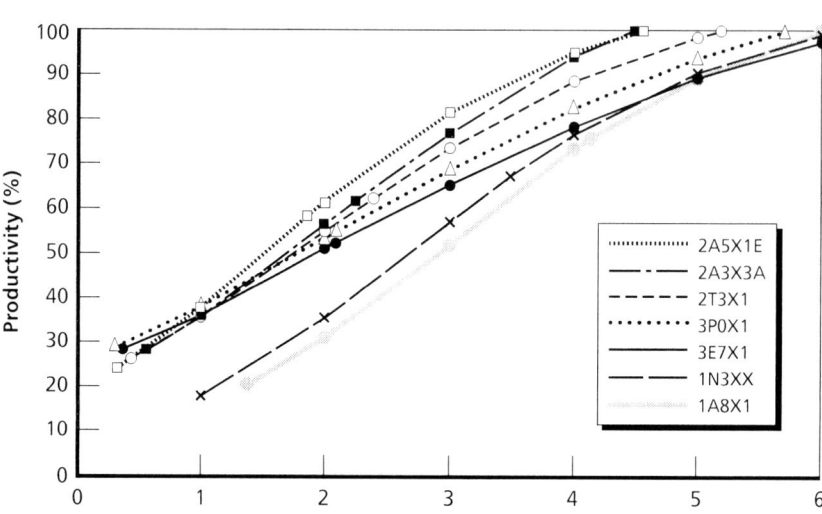

NOTE: 1A8X1 = Airborne Cryptologic Linguist; 1N3XX = Cryptologic Linguist;
2A3X3A = F-15 Maintenance; 2A5X1E = B-1/B-2 Maintenance; 2T3X1 = Special
Purpose Vehicle Maintenance; 3E7X1 = Fire Protection; 3POX1 = Security Forces.
RAND *MG555-3.6*

ness curves for these specialties are similar, we conclude that the rates of
learning in and out of IST are roughly the same. IST provides a good
foundation, but airmen continue to learn the language at approximately
the same rate in OJT. Because productive work is also accomplished
in OJT, it may make sense to do more of the learning in OJT. In order
to better understand whether IST reductions might have unexpected
consequences, we look at the effect of removing content based on the
SMEs' survey opinion.

This figure also shows that the five noncryptologic specialties all
start in the same general area of IST completion time and productivity,
25- to 30-percent. Among these, the two aircraft maintenance specialties (AFSCs 2A3X3A and 2A5X1E) start marginally lower but reach
fully productive status faster than the rest. During IST, these two specialties do not train on real equipment. Rather, their first exposure to
the actual weapon system is at their unit of assignment. These main-

tenance specialties depend on Field Training Detachments (FTDs) to provide hands-on, concentrated training. The FTD is, in essence, a hybrid between IST and OJT. One hypothesis for the rapid rise in effectiveness is the important effect of FTD and the ability to train on real systems.

Determining What Curriculum Elements Should Be Dropped or Added

Having developed the productivity curves, we next focused on the questions related to incremental changes to IST. Before discussing how we quantified these changes, we first look at the qualitative responses given in response to the questions of what should be added or removed from IST. In particular, this section considers response patterns within the open-ended response section in questions 19 and 20. We not only analyze specific suggestions given by respondents but also consider the implications of a blank or skipped response.

The survey did not specifically provide an option for respondents to indicate that they would prefer "no change" to IST. Instead the survey asked respondents to provide textual replies to questions on additions to and deletions from IST. In Figure 3.7, we see that, on average, only a small percentage of respondents, approximately 5 to 10 percent, wrote in responses that could be readily interpreted as "don't add" or "don't delete." This would suggest some level of dissatisfaction with the current training program.

Of note are the similar response rates of all but one specialty. Only Security Forces had fewer "don't change" responses, and they were significantly fewer. One interpretation of this result is that members of the Security Forces specialty are the least satisfied with the current IST curriculum.[9]

---

[9] We must strongly caveat this result. We are not saying that the Security Forces are dissatisfied with the current IST program. Rather, the survey data suggest that, among the seven AFSCs, the Security Forces may be the most dissatisfied with their current IST curriculum. Our survey was not directed to determine curriculum satisfaction, and so a more careful analysis should be performed in order to substantiate this inference.

Figure 3.7
Respondents Recommending Neither Increases Nor Decreases in IST

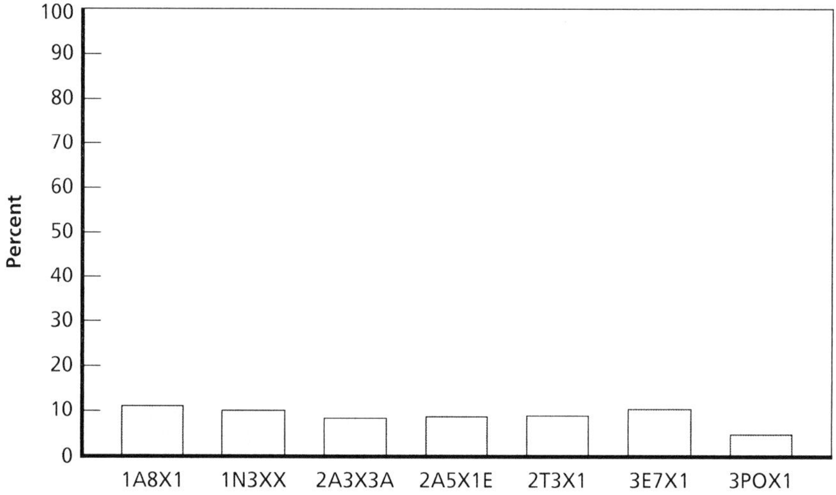

NOTE: 1A8X1 = Airborne Cryptologic Linguist; 1N3XX = Cryptologic Linguist;
2A3X3A = F-15 Maintenance; 2A5X1E = B-1/B-2 Maintenance; 2T3X1 = Special
Purpose Vehicle Maintenance; 3E7X1 = Fire Protection; 3POX1 = Security Forces.
RAND MG555-3.7

We also looked at the percentage of individuals who chose to skip one or both drop/add questions and those who wrote in "none" (Figure 3.8). We do not know for sure the reason why an individual did not respond to a particular question. It is possible some who had strong opinions about changes were frustrated because they were not given enough space to write out their opinions.[10] For this analysis, we assume that the individuals who skipped were, for the most part, satisfied and had no suggestions for changes.

If we compare the "drop" bars to the "add" bars, we note that in that in the majority of specialties, individuals skipped the "drop" question (or wrote in "none") more than the "add" question. In other words, while a significant portion may not have had an opinion regarding

---

[10] The survey allowed only one suggestion for "add" and one for "drop." The suggestion could be a course area, but there were only 256 characters for a response. Some respondents did comment on the lack of space to write out their ideas.

Figure 3.8
Additional Responses to the Drop/Add Question

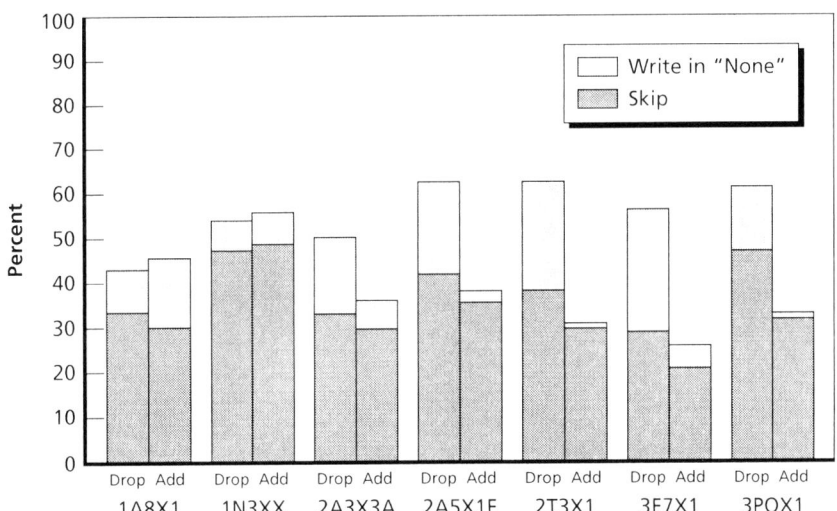

NOTE: 1A8X1 = Airborne Cryptologic Linguist; 1N3XX = Cryptologic Linguist;
2A3X3A = F-15 Maintenance; 2A5X1E = B-1/B-2 Maintenance; 2T3X1 = Special
Purpose Vehicle Maintenance; 3E7X1 = Fire Protection; 3POX1 = Security Forces.
RAND MG555-3.8

what to drop from the curriculum, they definitely had ideas on what
to add.

For the "add nothing" group (the top of the right-hand bar of
each set), the individuals in the two longer technical training courses
(AFSCs 1A8X1 and 1N3XX) were much more likely to suggest "no
addition to technical training" than those in a specialty with a shorter
training duration. The cryptologist specialties may not have the same
enthusiasm to add content to IST programs because the associated
courses are already over a year in length. The data are inconclusive,
though, for dropping some of the course content.

Discussion of "Suggestion" Responses

It is also worthwhile to look at the responses from the opposite per-
spective—comparing the patterns of respondents who did have suggestions for additions and deletions. Figure 3.9 indicates the percentage of

Figure 3.9
Respondents Who Suggested Both Deleting and Adding

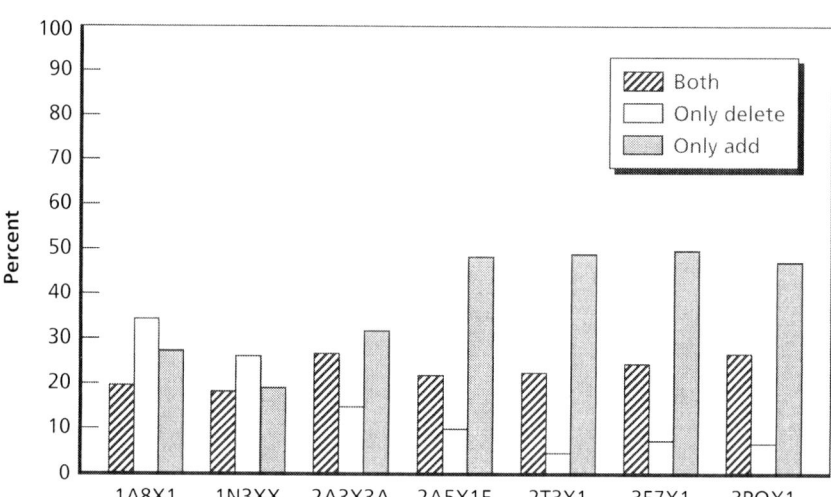

NOTE: 1A8X1 = Airborne Cryptologic Linguist; 1N3XX = Cryptologic Linguist;
2A3X3A = F-15 Maintenance; 2A5X1E = B-1/B-2 Maintenance; 2T3X1 = Special
Purpose Vehicle Maintenance; 3E7X1 = Fire Protection; 3POX1 = Security Forces.
RAND MG555-3.9

respondents who had a suggestion for both deleting and adding course
elements alongside those who provided only a suggestion for something
to delete and those who provided only a suggestion for something to
add. The group of respondents who provided both an add suggestion
and a delete suggestion are relatively consistent (19–28 percent) across
all seven specialties. Respondents who provided both a "drop" and an
"add" suggestion highlight the potential for curriculum trades.

The message is similar to that of Figures 3.7 and 3.8. The respondents in specialties with longer IST were more likely to have only suggestions for something to delete; the respondents in the shorter IST
programs were overwhelmingly more likely to have only suggestions
for something to add. In Figure 3.10, we compare the percentage of
respondents suggesting something to add versus those who specifically
wrote "No, don't add anything" in Question 19. In the programs with
shorter IST, respondents were strongly in favor of adding; very few
respondents specifically suggested that nothing should be added (most

Figure 3.10
Respondents Who Suggested Adding Compared with Those Who
Suggested Not Adding

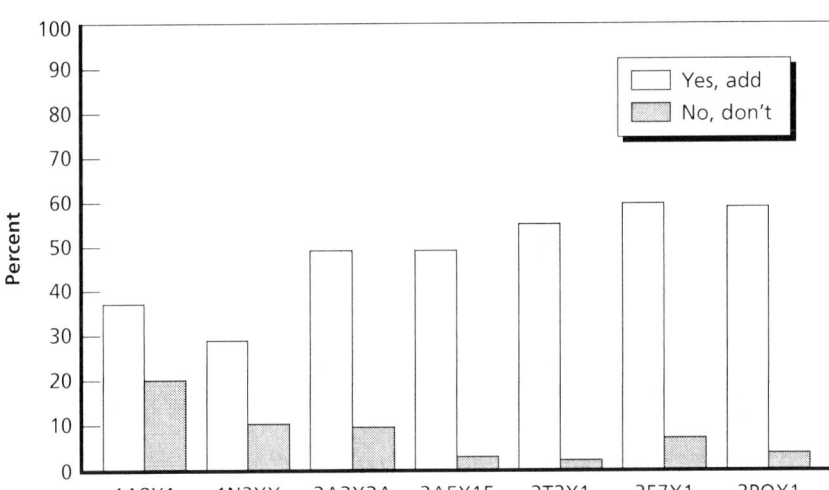

NOTE: 1A8X1 = Airborne Cryptologic Linguist; 1N3XX = Cryptologic Linguist;
2A3X3A = F-15 Maintenance; 2A5X1E = B-1/B-2 Maintenance; 2T3X1 = Special
Purpose Vehicle Maintenance; 3E7X1 = Fire Protection; 3POX1 = Security Forces.
RAND MG555-3.10

notably the 2A5X1E, 2T3X1, 3E7X1, and 3P0X1 specialties). In the
longer IST programs, that result is not as clear. A much smaller percentage of respondents suggested an addition, and the difference between
those who suggested adding and those who insisted that no additions
be made is considerably smaller.

Similarly, Figure 3.11 compares the percentage of respondents
answering with a "delete" suggestion versus those who wrote, "No,
don't delete anything." In this case, the results are much less striking.
Respondents in the specialties with the longest IST courses were more
than twice as likely to suggest something to delete as to request no
deletion at all. In the five specialties with shorter IST, the percentage
of respondents suggesting a deletion is similar to those suggesting that
no deletions be made.

Figure 3.11
Respondents Who Suggested Dropping Compared with Those Who
Suggested Not Dropping

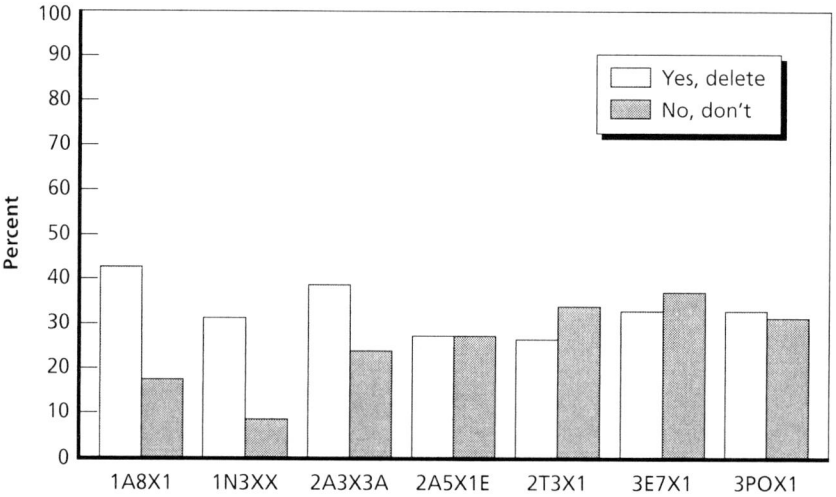

NOTE: 1A8X1 = Airborne Cryptologic Linguist; 1N3XX = Cryptologic Linguist;
2A3X3A = F-15 Maintenance; 2A5X1E = B-1/B-2 Maintenance; 2T3X1 = Special
Purpose Vehicle Maintenance; 3E7X1 = Fire Protection; 3POX1 = Security Forces.
RAND MG555-3.11

A final observation comes from an examination of the effect on
productivity for dropping course content. Respondents had no difficulty associating an increase in training with an increase in productivity. The increase in productivity with an increase in the length of
technical training implied a value-added block of instruction. The
responses for dropping content were much different. When asked about
the productivity effect at IST graduation, the majority of respondents
who suggested dropping content answered that there would be zero
effect. This implies that the respondents felt some course element had
no value added in terms of productivity. Figure 3.12 shows the percent of respondents who suggested dropping content with no loss
in productivity—in other words, that some training time had been
wasted.

Figure 3.12
Percentage of Respondents Who Suggested Drops from IST with No Effect

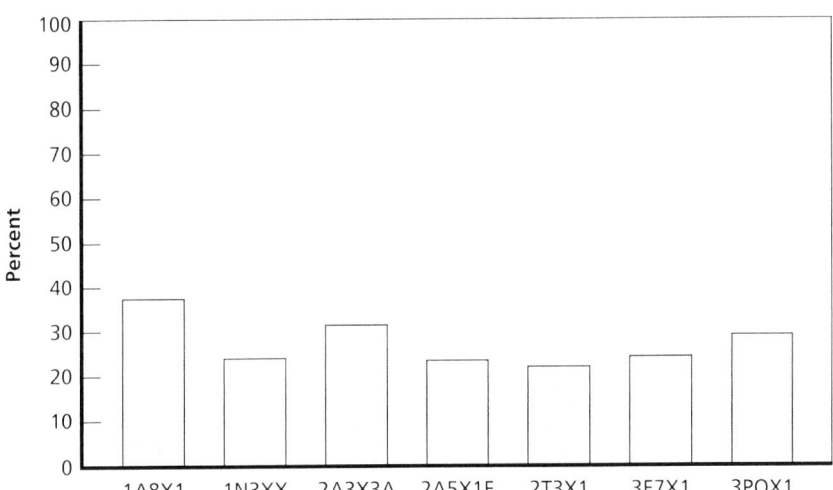

NOTE: 1A8X1 = Airborne Cryptologic Linguist; 1N3XX = Cryptologic Linguist;
2A3X3A = F-15 Maintenance; 2A5X1E = B-1/B-2 Maintenance; 2T3X1 = Special
Purpose Vehicle Maintenance; 3E7X1 = Fire Protection; 3POX1 = Security Forces.
RAND MG555-3.12

To better understand these results, it is helpful to consider some examples within the specialties. In the case of AFSC 1A8X1, there is some redundancy in the program, which functions as a screening tool.[11] An individual in this specialty might well see this redundancy as unnecessary. Individuals in AFSCs 2A3X3A and 2A5X1E learn basic maintenance until they actually arrive on base and start to receive training on their specific aircraft. The nonspecific parts of the course may appear unnecessary. AFSC 3P0X1 has numerous roles in the Air Force (base defense, law enforcement, nuclear security, and—more recently—deep battle engagement). Individuals receive some training for each of these roles even though their initial assignment will not utilize all the training elements.

---

[11] The relationship of redundancy to screening is the following: Some of the instruction is done early on and quickly to find people who are unable to grasp it. The instruction is then revisited later in the program at the depth required for proficiency.

Summarizing the Qualitative Data

We conclude this section with some general observations based on the above qualitative analysis. Table 3.4 summarizes some of the indicators from the previous figures by specialty. The column "Write-in Comments" refers to indicators from respondents' written comments. The column "Add vs. Drop Comparison" refers to whether the preponderance of responses favored adding or dropping content. The analysis above suggests that respondents in AFSCs 1A8X1 and 1N3XX recommend dropping curriculum content, with those in AFSC 1A8X1 providing strong evidence in their write-in comments. Results for AFSCs 2T3X1, 3E7X1, and 3P0X1 suggest adding to their curricula.

A detailed analysis of the written comments is found in Appendix G.

Table 3.4
Summary of Specific Survey Responses to Drop/Add Questions

AFSC	Specialty Title	Write-in Comments	Add vs. Drop Comparison
1A8X1	Airborne Cryptologic Linguist	Don't add	Drop
1N3XX	Cryptologic Linguist		Drop
2T3X1	Special Purpose Vehicle Maintenance		
2A3X3A	F-15 Tactical Aircraft Maintenance		
2A5X1E	B-1/B-2 Aerospace Maintenance	Don't drop	Add
3P0X1	Security Forces	Don't drop	Add
3E7X1	Fire Protection		Add

Developing Incremental Change Functions

We now turn our attention to a quantitative analysis of the drop/add responses. In each case, respondents were asked to name a skill to be dropped or added, then to estimate the length of the training content and the resulting change in productivity.

Figure 3.13 represents most of the responses for AFSC 2T3X1. The x-axis represents the suggested change in the number of training days. The y-axis represents the change in the effectiveness percentage at the end of IST. The figure shows curves in only two quadrants: the upper right, in which adding days increases final productivity, and the lower left, in which subtracting days decreases productivity.[12] (Some responses were off the chart.) We used these data to fit curves, one for each quadrant, using a second-order polynomial model. These curves show an expected decreasing return to scale in both quadrants. We also estimated the upper and lower bound of the 95-percent confi-

Figure 3.13
Fitting a Curve to the Drop/Add Questions, AFSC 2T3X1

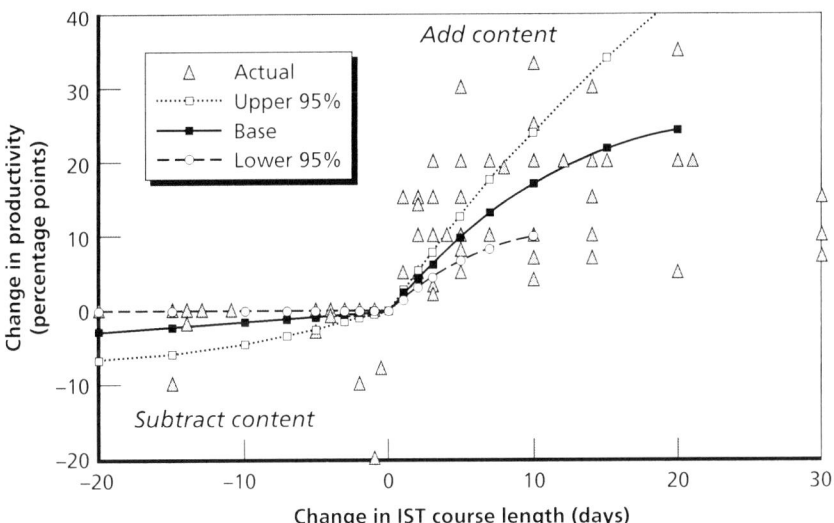

---

[12] As noted earlier, we did have a small number of responses that added days and reduced productivity or subtracted days and added productivity (typically 0–3 respondents per specialty). Because it was unclear whether the respondents misunderstood the question or did not know how to answer the question properly, these observations were dropped.

dence interval. We fit curves for each of the specialties achieving statistically significant fits.[13]

The computed lines likely overestimate the effect of changes to the curriculum. This is due to the fact that we averaged the individuals' "pet rocks," thereby creating a consistent bias in the predicted curve. For example, one person suggested adding one day of "A" for an increase in productivity of 10 percent and another suggested adding one day of "B" for an increase of 15 percent. If those were the only two responses, we would develop a point estimate relating a one-day increase in training to a 12.5-percent increase in effectiveness. But if the person who said "A" was asked about the effect of adding "B", he might say that only a 2-percent change in productivity would occur. Likewise, the person who suggested "B" might think that "A" improves productivity by only 1 percent. Then, on average, we would have a 6-percent change for adding "A" and an 8-percent change for adding "B." Therefore, a one-day change in the course would increase effectiveness by either 6 percent or 8 percent, depending on what was added, not by the 12.5-percent that is based on the information available for our evaluation.

An alternative way of approaching the add/delete analysis would have been to determine the top 10 items of content to add and the top 10 items of content to drop, as well as the length of training time for each item as determined by experts. We would have then asked each individual to determine the effect on productivity at the end of IST that adding or deleting that item would have. If we had used this approach, we could have estimated the change in productivity for adding or subtracting course elements with a greater degree of accuracy.

Despite these limitations, we wanted to use these data and the derived curves to help estimate the effects of changing the length of IST. Note again that there is much greater variation in the "add" data than in the "drop" data. In particular, variance in the data and confidence intervals for adding are three to four times greater than the vari-

---

[13] All the add functions were significant at the 90-percent level. The drop curves for AFSCs 2T3X1 and 2A3X3A were not statistically significant at the 90-percent level, primarily because of the small number of responses.

ance in the data for dropping. Additionally, the loss of effectiveness for dropping is generally quite small, with a large number of the "drop" responses suggesting that reducing some course content would cause no loss in productivity. Because of this, it made little difference which curve we used to represent the "drop" data. Therefore, we used the median line. On the contrary, the wide variation in the "add" data made the selection of a curve to represent these data more sensitive. We wanted to recommend additions to IST only when there was strong supporting evidence. We also wished to minimize the general overestimation of the effect of adding course length. For these reasons, we used the lower limit of the confidence interval as our estimate of the impact of adding to course length. This also allowed us to appeal to an a fortiori argument based on these conservative assumptions that only strengthens any recommendation to add to course length.

Figure 3.14 shows the fitted curves for adding content for each of the seven specialties. Comparing the lower limits of the confidence interval, we find that AFSCs 1A8X1 and 1N3XX show the least increase in productivity for increases in course content, whereas AFSC 3P0X1 shows the greatest payoff for increasing course content. This is consistent with the results we have already discussed.

A simpler measure that captures the desirability of adding course content is the average change in productivity at graduation across all observations. Table 3.5 shows these values ordered from least to greatest. Larger values suggest greater effects in productivity from adding days. Similarly, small changes in the average productivity suggest that less is to be gained by increasing the course content. Similar to the curves in Figure 3.14, AFSCs 1N3XX and 1A8X1 have the smallest change, and AFSC 3P0X1 shows the largest average change.

For dropping content (decreasing course length), we wanted to eliminate instruction that has the smallest impact on productivity.

Figure 3.14
Incremental "Add" Functions

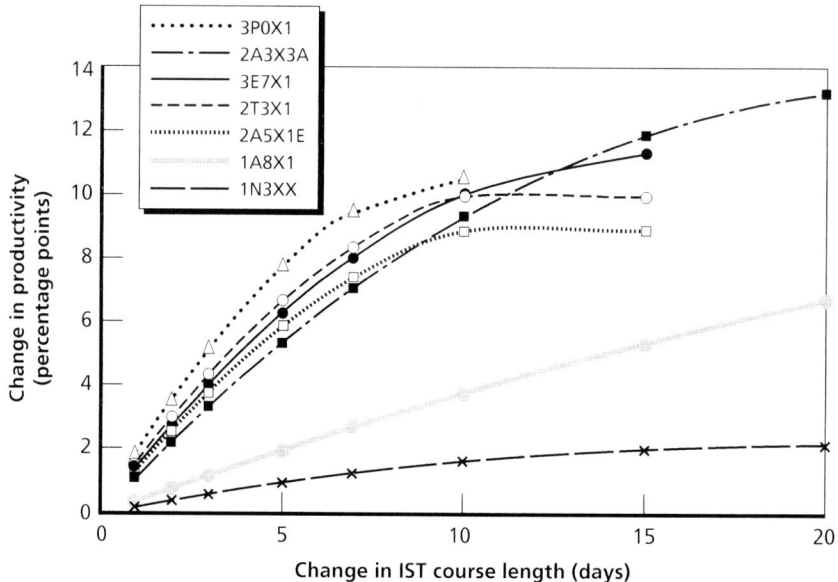

NOTE: 1A8X1 = Airborne Cryptologic Linguist; 1N3XX = Cryptologic Linguist;
2A3X3A = F-15 Maintenance; 2A5X1E = B-1/B-2 Maintenance; 2T3X1 = Special
Purpose Vehicle Maintenance; 3E7X1 = Fire Protection; 3POX1 = Security Forces.
RAND MG555-3.14

Table 3.5
Average Change in Productivity for Adding One Day in IST

Specialty	Average Change in Productivity for One Additional Day of IST
1A8X1: Airborne Cryptologic Linguist	0.35
1N3XX: Cryptologic Linguist	0.40
2A3X3A: F15 Maintenance	0.53
2T3X1: Special Purpose Vehicle Maintenance	0.91
3E7X1: Fire Protection	0.96
2A5X1E: B1/B2 Maintenance	0.96
3POX1: Security Forces	1.64

Figure 3.15 shows the fitted lines for all seven AFSCs. The two cryptologic specialties, AFSCs 1A8X1 and 1N3XX, have curves essentially lying on the negative x-axis. This means that decreasing IST would have little impact on the productivity level at IST graduation. On the other hand, even a very small decrease in content for AFSC 3E7X1 would have a huge impact in the productivity of IST graduates. This result is understandable given that students must reach a well-specified skill threshold—national certification to fight fires—before graduation. Any reduction in training would result in the loss of their firefighting certification and a need for more training at the unit before they could begin contributing to the mission.

If we look at the average change in productivity (Table 3.6), we see results similar to the fitted curves. For deletions from IST, it is desirable to have the smallest average value indicating the least impact

Figure 3.15
Incremental Drop Functions

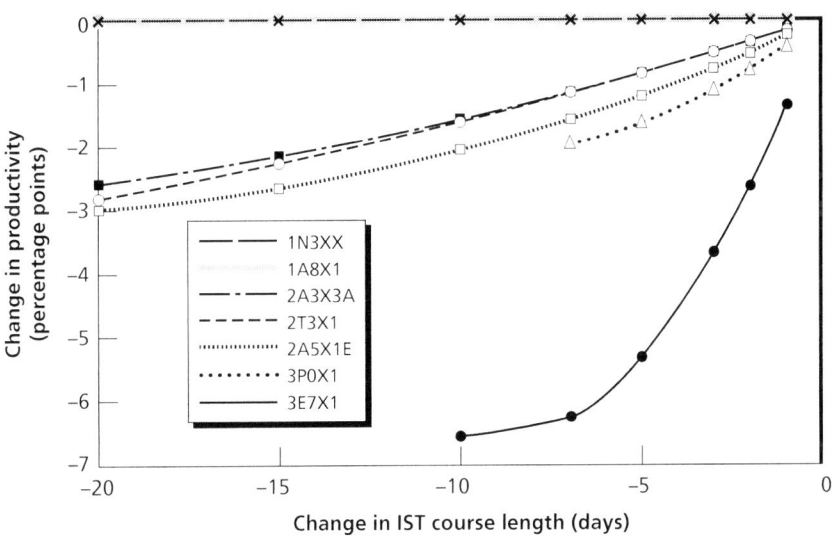

NOTE: 1A8X1 = Airborne Cryptologic Linguist; 1N3XX = Cryptologic Linguist; 2A3X3A = F-15 Maintenance; 2A5X1E = B-1/B-2 Maintenance; 2T3X1 = Special Purpose Vehicle Maintenance; 3E7X1 = Fire Protection; 3POX1 = Security Forces.
RAND MG555-3.15

Table 3.6
Average Change in Productivity for Dropping One Day in IST

AFSC	Average Change in Productivity for One Less Day of IST
1N3XX: Cryptologic Linguist	0.06
2A3X3A: F15 Maintenance	0.11
1A8X1: Airborne Cryptologic Linguist	0.19
2A5X1E: B1/B2 Maintenance	0.20
3P0X1: Security Forces	0.41
2T3X1: Special Purpose Vehicle Maintenance	0.47
3E7X1: Fire Protection	1.89

on graduation effectiveness for reductions in IST. Once again, AFSC 3E7X1, Fire Protection, has a very large negative impact on productivity for one less day of training.

Based on these results, we now expand our summary in Table 3.7. The estimated incremental change functions (drop/add curves) reflect the rate of change in productivity for increases or decreases in course length and, by extension, course content. The drop/add curves suggest that AFSCs 1A8X1 and 1N3XX might be candidates for decreas-

Table 3.7
Expanded Summary of Specific Survey Responses to Drop/Add Questions

AFSC	Specialty Title	Write-in Comments	Add vs. Drop Comparison	Incremental Change Functions
1A8X1	Airborne Cryptologic Linguist	Don't add	Drop	Decrease
1N3XX	Cryptologic Linguist		Drop	Decrease
2T3X1	Special Purpose Vehicle Maintenance			Increase
2A3X3A	F-15 Tactical Aircraft Maintenance			Increase
2A5X1E	B-1/B-2 Aerospace Maintenance	Don't Drop	Add	Increase
3P0X1	Security Forces	Don't drop	Add	Increase
3E7X1	Fire Protection		Add	Increase Don't decrease

ing course length. The results from the other five AFSCs suggest some payoff for increased course content. AFSC 3E7X1 shows a strong negative effect for any reduction in course length.

The preceding analysis has focused on specific survey insights. We have not yet examined these possible changes in the light of current productivity at the unit level. In the next chapter, we use the productivity curves, the change functions, and the calculated costs of a specialty to determine optimal productivity for a given cost.

CHAPTER FOUR

Calculating Total Cost

Overview

A key component of a cost-benefit analysis is measuring the total cost of each alternative. To cost out each of the seven specialties, we included the first order costs associated with direct salaries as well as IST and OJT. We relied on a combination of sources. Our primary source was AFI 65-503 (Secretary of the Air Force, 1994b), attachments 17-1, 18-1a, and 19-2. We also used the Air Education and Training Command, Directorate of Operation's (AETC/DOR's) Technical Training Cost Model for advanced training costs. When needed, we used the weighted inflation rates of the Deputy Assistant Secretary of the Air Force for Cost and Ecomonics (SAF/FMC) to convert to a common year, 2005. We do not claim that these costs represent the detail level of a POM cost estimate. However, they do include the primary cost elements, they all come from the same sources, and they all have the same assumptions applied. Although it is important and desirable for the costs to be as complete and accurate as possible, it is far more important in a cost-benefit analysis that the costs are consistent relative to each other.

Productive Activity and Human Capital Development Costs

In developing pay-related costs, we used the composite pay methodology described in AFI 65-503. Composite pay is a function of basic pay for grade and year of service, retired pay accrual (27.5-percent), a flat medical health care accrual, basic allowance for housing based on grade, average incentive and special pays (to include aircrew) based on grade, and a miscellaneous pay based on grade. The miscellaneous pay includes subsistence, family separation allowance, separation payments, employer's contribution of the social security tax, overseas station allowances, death gratuities, reenlistment bonuses, special duty assignment pay, clothing allowances, and unemployment compensation. We did not include permanent change of station costs.

We divided total pay-related costs into two categories, the cost of productive activity and the cost of HCD. The cost of productive activity is defined as composite pay times the productivity level of the individual. HCD cost is defined as the composite pay times 1 minus the individual's productivity level. Our definition of HCD cost is the common way to measure HCD in the human capital accounting literature (Flamholtz, 1999).

Training Costs

Training costs were taken from AFI 65-503, Table A18-1a, which is compiled from AETC's Training Cost Report database. We also included BMT in the initial skills cost. We used AFI 65-503, Table A17-1, for the accession and recruiting costs.

For increased course lengths, we proportionally increased the cost of IST based on its initial cost and length. We did the analogous decrease in cost for decreases in IST course length.

For advanced training costs, we determined an historical average cost per person for each AFSC. There are over 1,000 advanced skills and professional military education (PME) courses historically taken by individuals in our seven AFSCs. We used the actual length of each course and the historical percentage of personnel by experience-level attending the course to develop an approximate cost. We derived course attendance from personnel data.[1] Table 4.1 shows the non-PME

---

[1] Unfortunately, a perfect match between personnel records and the advanced training course data was not possible. Many course codes in the personnel data could not be directly

advanced training that was estimated for AFSC 2T3X1. Column 1, the level, is the approximate skill level of an individual receiving the training; column 2 is the historical percentage of individuals at that level who received the training.

For advanced training costs, we used published AETC data and added a factor for per diem and travel. Total advanced training costs were small compared to other costs and our results were not sensitive to changes in the associated cost factors.

On-the-Job Training Costs

For OJT costs, we used the composite pay times 1 minus the productivity level. In other words, OJT cost is defined as the cost of paying an individual's salary for non-productive time. This is based on the assumption that the less effective an individual is, the more time the individual spends learning or in training. We were not able to include the cost of the supervisors' time or any increased maintenance costs as unit experience levels dropped (e.g., increases in poorer quality work or mistakes such as broken equipment). We did adjust the OJT learning rate if 5–7 level manning fell below authorized strength. Essentially, we used a factor that slowed the rate of OJT attainment by the percentage of the manning shortfall. In the steady state analysis, this rarely occurs.

---

matched with codes in the course list. Our first attempt resulted in the ability to match codes and identify courses for only about 50-percent of the entries in the personnel records. Some common errors in entering course codes into the training records included mistaking similar characters, such as numeric 0 and alphabetic O. After resolving the common errors and examining some course descriptors, it was possible to greatly increase the number that matched. Eventually, we were able to match about 80-percent of the codes. Since most of the unmatched codes were for courses that were not widely attended, 80-percent of the codes accounted for 95-percent of the training days.

Table 4.1
Example of Additional Training in AFSC 2T3X1

Level	%	Length (days)	Title
3	82	50	Interservice Mechanic Apprentice Course
7/9	41	20	Vehicle Maintenance Superintendent
3/5	39	3	Air Force Training Course
5	13	11	A/S32 P-23 Fire Truck Organizational/Intermediate (O/I) Maintenance
3/5	10	12	Diesel Engine Maintenance
3/5	10	8	Landoll Deicer O/I Maintenance
5	9	10	Diesel Engine Maintenance
3/5	8	20	Vehicle Diagnostic Test Equipment and Electrical Systems
3/5	7	10	Steering, Suspension, Wheel Alignment, and Antilock Brakes
3/5	7	5	Vehicle Air Conditioning Systems
5	7	13	A/S 32p-19 Fire Truck O/I Maintenance
5	7	12	Oshkosh R-11 Refueler O/I Maintenance
3/5	6	5	FMC Halvorson 25k Cargo Loader Maintenance Course
5	6	14	System and Electronic Inc 60k Loader O/I Maintenance
5	6	12	Automatic Transmission/Transaxle Maintenance
5	5	10	Systems and Electronics 60k Loader O/I Maintenance
5	5	12	Kovatch R-11 Refueler O/I Maintenance
3/5	4	64	Vehicle Body Mechanics (USAF)
5	4	2	Air Force Technical Order System
3/5	3	5	Power Steering, Antilock Brake and Supplemental Restraint Systems
3/5	3	8	Vehicle Test Equipment/Electrical System Analysis
5/7	3	22	Teaching Internship
5/7	3	4	Principles of Instructional Systems Development
5	3	10	Vehicle Maintenance Control and Analysis Craftsman
5	3	5	Vehicle Test Equipment
5/7	2	5	Objectives and Tests
5/7	2	4	Space Corp 40k A/Sh-6 Cargo Loader O/I Maintenance
3	2	7	Southwest Mobile 25k Cargo Loader Maintenance
3	2	10	Systems And Electronics 60k Loader O/I Maintenance

Table 4.1—continued

Level	%	Length (days)	Title
5	2	9	Southwest Mobile 25k Loader O/I Maintenance
5	2	10	Tri-State R-12 Hydrant Hose Truck O/I Maintenance
5	2	5	Power Steering and Power Brakes Maintenance
7	3	7	Training Supervisor
7	2	25	Basic Instructor Course
7	1	3	Sorts Data Handler (USAF)
7	1	13	A/S32 P-15 Fire Truck O/I Maintenance
7	1	3	Environmental Compliance Assessment and Management Program
7	1	3	Pollution Prevention Program Operations and Management

CHAPTER FIVE

Cost-Effective Course Lengths

In this chapter, we integrate the analysis of the previous chapters and demonstrate ways in which the Air Force can determine the length of IST that will minimize training costs, maximize productivity, or minimize force strength. Cost trade-offs are, of course, a primary concern for the Air Force. However, overall productivity has a significant impact on the quality of the force and its ability to perform specific missions. Force size also becomes important particularly in times of force reductions, as are projected over the next five to six years. Furthermore, these measures are not independent, and so it is also important to understand the interactions among them.

We begin this chapter by answering the question, How can we calculate the full cost of human capital development? Once this question is answered, we can use the same methodology to determine whether adding or subtracting from IST course lengths will increase or decrease costs. This leads us to discover the IST/OJT split that has the most impact on our outcome measures. Finally, we focus on each of the three measures of performance: cost, productivity, and force strength.

The Full Cost of Human Capital Development

Figure 5.1 depicts the population of AFSC 2T3X1[1] by YOS and skill level. The steady state force is characterized by constant attri-

---

[1] We use AFSC 2T3X1 as our example case throughout this chapter.

tion[2] and reenlistment rates. The increase in the force that occurs between YOS 9 and 14 is the result of crossflows in and out of other specialties.

Figure 5.2 depicts the previously scaled productivity curve (Figure 3.5), now represented in discrete years of service.

To calculate total productivity, we take the productivity at each YOS and multiply it by the population of that YOS and by the percentage of airmen who are not in additional training in that YOS. We then sum the individual year of service productivity values into a single *productivity index* for the specialty (see equation below). The productivity index is the number of fully qualified journeymen who would produce the same outputs as the specialty's given mix of journeymen and OJT

Figure 5.1
Steady State Population by YOS and Skill Level, AFSC 2T3X1

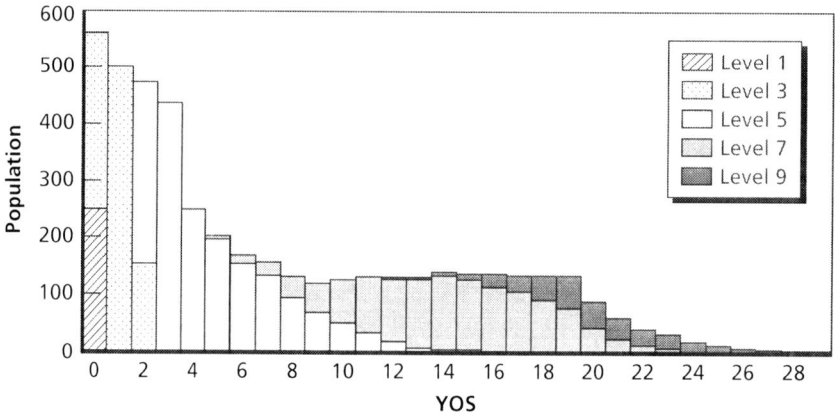

NOTE: YOS 1 population includes projected grads. Individual steady state skill level totals are not the same as authorization totals. Total steady state skill level values equal total authorizations.
RAND *MG555-5.1*

---

[2] Attrition varies by YOS but not year to year. That is, a 2005 population will experience the same rates as the projected 2010 population.

Figure 5.2
Discrete Productivity Values by YOS, AFSC 2T3X1

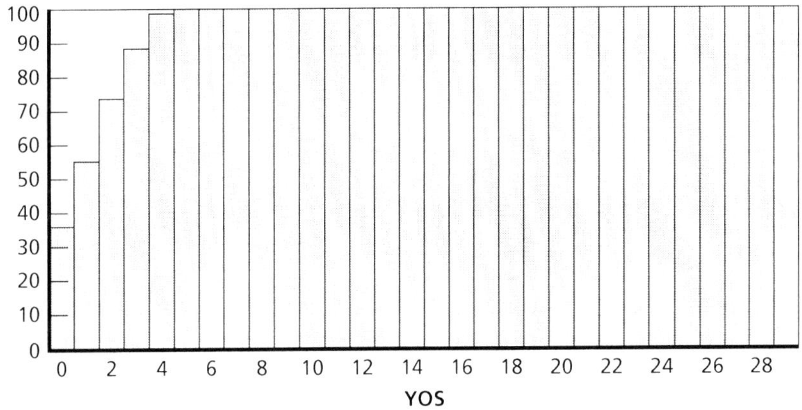

trainees. Figure 5.3 shows the population by YOS and the corresponding productivity values used to compute the productivity of the force.[3]

$$T = \sum_{i=0}^{29} P_i \cdot W_i \cdot N_i$$

where
T = sum of the productivity of the force over all years of service
(0 to 29)
P_i = population of year of service i
W_i = productivity for one person in year of service i
N_i = percentage of airmen who are not in additional training for year
of service i.

Likewise, we calculated the pay costs associated with each individual year of service separated into costs of productive activity and OJT.
Total HCD costs would include costs associated with OJT, recruiting,

---

[3] Not shown graphically is the third factor, the percentage of airmen who are not in additional training.

Figure 5.3
Calculating Total Productivity

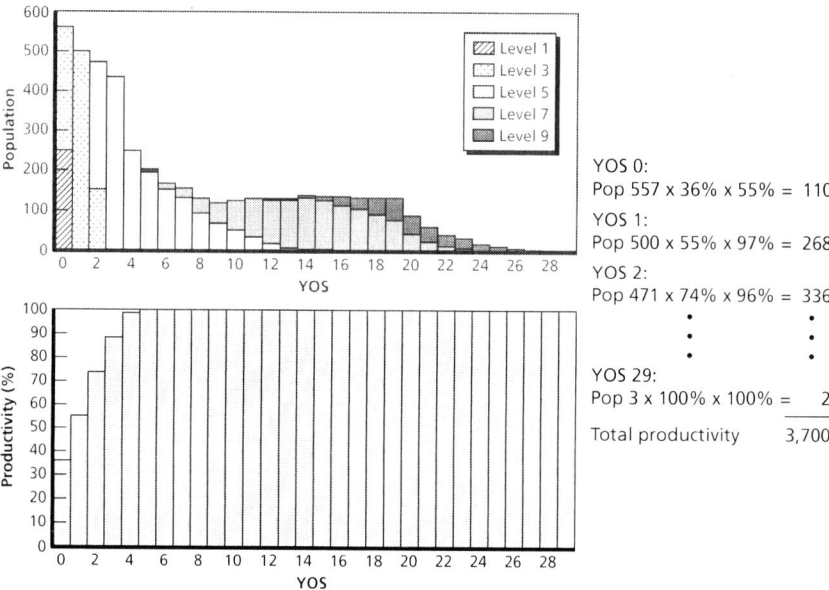

YOS 0:
Pop 557 x 36% x 55% = 110

YOS 1:
Pop 500 x 55% x 97% = 268

YOS 2:
Pop 471 x 74% x 96% = 336

•   •
•   •
•   •

YOS 29:
Pop 3 x 100% x 100% =   2
                    _____
Total productivity   3,700

accessing, and initial skills training. Summing all costs give us the total cost of the AFSC 2T3X1 force (Table 5.1).

In this case, the productivity of a steady state force of 4,357 individuals is 3,700 fully effective equivalents. Dividing the total costs, $304.6 million, by the total productivity, we get the value of $82,330—the cost for generating one unit of productivity. This gives us the key methodology needed to determine the cost of human capital development.

Reducing Human Development Costs

The next question is, Can we reduce the cost of generating a unit of productivity by a different division between IST and OJT? In Figure 5.4, we take our scaled productivity curves and apply the results of the on incremental change function. Although they are difficult to discern

Table 5.1
Cost Measures, AFSC 2T3X1

Category	$ Million	Other Measures	Value
Productive activity	188.4	Actual force size	4,357
OJT	23.2		
Recruit/access	4.3	Total production	3,700
Initial skills training	16.7		
Advanced training	3.5		
Retirement accrual	68.6		
Total cost	304.6[a]	Total cost/total production	$82,330

[a]Numbers do not sum exactly because of rounding.

Figure 5.4
Effect of IST Changes on Productivity Curves, AFSC 2T3X1

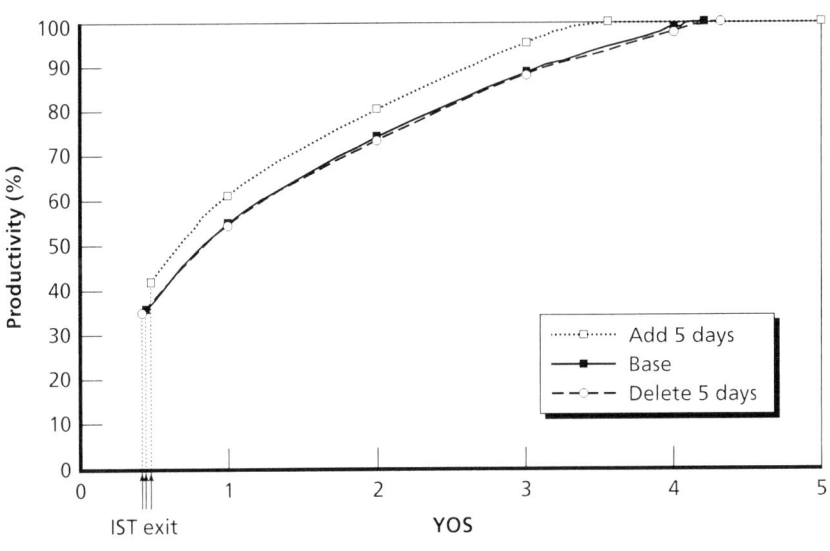

RAND MG555-5.4

on the figure, each of the three lines has a different YOS starting point. The base case starts at approximately 0.45 years of service. The "Add 5 Days" line starts at 0.47 years, and the "Delete 5 Days" line starts at 0.43 years.

As noted earlier, these effects are exaggerated because of the cumulative effect of individual recommendations to add elements, even when we discount this effect by using the lower 95-percent confidence curve.

Cost, Productivity, and End Strength

We used a steady state force structure to examine the effect of changing IST length for each measure. First, we held work force strength (the total size of the working force) constant, then we held productivity constant, and finally we held the total cost of the force constant.

Constant End Strength

In Figure 5.5, work force strength is held constant while productivity and cost are allowed to vary. In this chart, the horizontal axis shows

Figure 5.5
Steady State Results (Maintain Work Force), AFSC 2T3X1

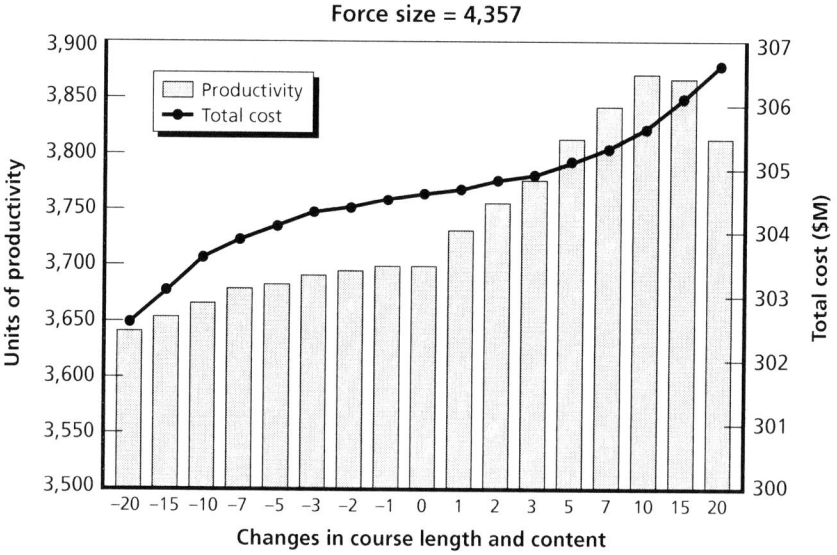

changes in course length (and by extension, content). The bars correspond to units of productivity and are measured using the vertical axis on the left. The curve corresponds to the total cost (in millions of dollars) using the vertical axis on the right.

The base case represents no change in course length and is the 0 point on the x-axis of the chart. As expected, holding work force strength constant, any increase or decrease in course length has a corresponding effect on cost. Additionally, decreases in course length decrease productivity; increases in course length, up to ten additional days, increase productivity. Increases beyond ten days do not increase productivity because the man-years lost to training are not offset by the increase in individual productivity. Another way to view this is that a 0.3-percent change in cost ($304.5 million to $305.5 million) results in a 5.0-percent change in productivity (3,700 to 3,850)—an order of magnitude increase.

Continuing the analysis, we then determine the cost per unit of productivity (Figure 5.6). Each bar is the cost per unit of productivity and is the result of dividing the total costs by the total productivity at each point. (In this graph, a lower value is better.) From the graph, we see that the cost per unit of productivity decreases as we move from the left (large decreases in course length) to the right (large increases in course length). At ten additional days of course length, the direction changes, and the costs increase at fifteen days of increased instruction. The shaded region of the chart represents the 95-percent confidence interval for the base case productivity curves. We ran the model for the lower and the upper limit of the confidence interval. The difference between them was $2,700 per unit of productivity ($1,350 above and below the mean).[4] Thus, cases with costs within this range are not statistically distinguishable from the base case. Alternatively, cases with costs outside this range can be differentiated statistically from the base case.

_____

[4] At the baseline value of no change to course length, the cost per unit of productivity ranged from $81,100 to $83,800.

Figure 5.6
Steady State Results (Cost/Benefit Curve), AFSC 2T3X1

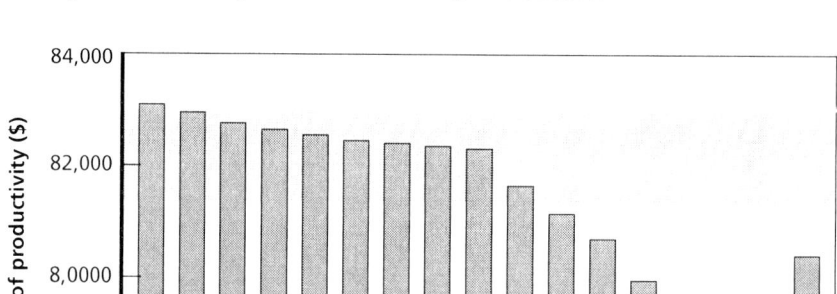

NOTE: Differences must be greater than $1,350 per unit of productivity for statistical significance.
RAND *MG555-5.6*

We conclude that, for a constant career field size, the Air Force could realize a cost per unit of productivity decrease by increasing the length of IST training for the 2T3X1 career field. Although our estimates of cost and productivity are approximate, the $3,500 difference in cost per unit of productivity for the addition of ten days of IST shown in this analysis is evidence that additional days of training should be considered. Furthermore, the ratios of cost to productivity shown above are also very useful in examining the trade-offs when either productivity or cost is held constant. In particular, the corresponding IST course lengths in the "maintain productivity" and "maintain cost" charts (Figures 5.7 and 5.8) that match the "maintain work force" chart will yield the same cost per unit of productivity.

Figure 5.7
Steady State Results (Maintain Productivity), AFSC 2T3X1

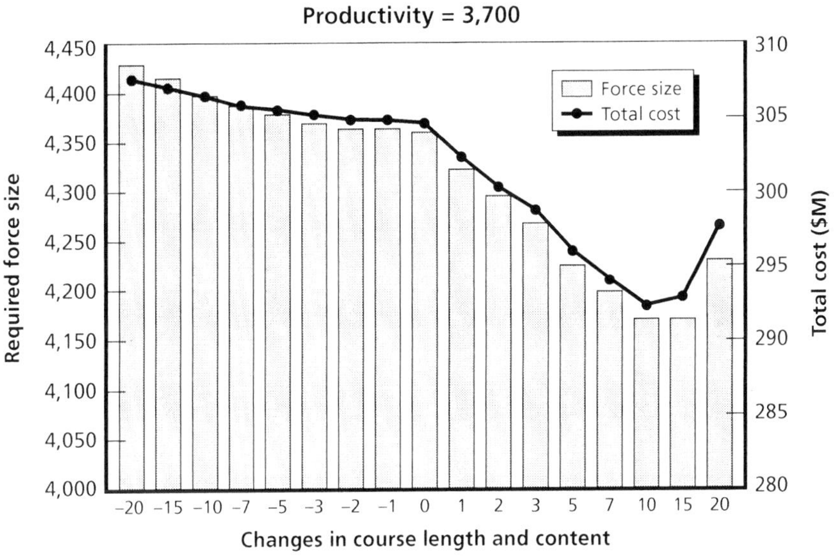

Productivity = 3,700

RAND *MG555-5.7*

Constant Productivity

In the "maintain productivity" case (Figure 5.7), we hold productivity constant at 3,700 equivalent fully effective workers and allow the work force size and total cost to vary.[5] In this situation, the increase in training costs (resulting from a longer IST course) is offset by reducing the work force size (manpower). This kind of trade-off is most significant at a time when overall force strength reductions are desired. This chart explores cases where lower cost and force strength may be attainable while keeping aggregate mission effectiveness (productivity) constant. Thus, an equally productive workforce is considerably smaller than the base case when ten days of IST are added, and is

_____

[5] We realize that in some cases, mission effectiveness requires having at least a minimum number of specialists to fill all the necessary positions. In these cases, the former trade between cost and productivity with fixed end strength will be more significant.

Figure 5.8
Steady State Results (Maintain Cost), AFSC 2T3X1

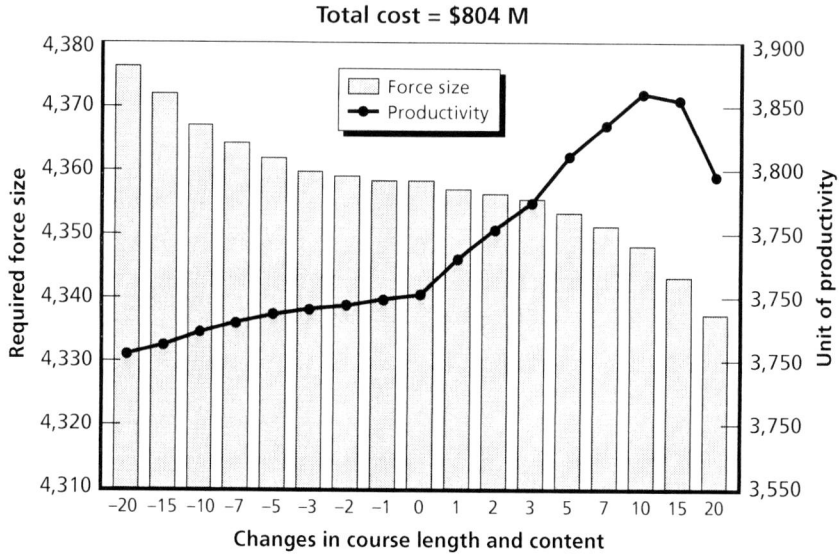

RAND MG555-5.8

therefore considerably cheaper. If we hold productivity at 3,700 equivalent fully effective workers and increase the course length by ten days, we see a reduction in force size from 4,350 personnel to approximately 4,200 personnel and a decrease in cost from $305 million to $293 million. The increase in productivity coming out of initial skills training (during the first five years) makes up for the loss of productivity associated with a smaller work force. This result is very significant. Increasing IST course length does increase IST costs, but the Air Force can maintain the same aggregate force effectiveness (productivity) with an overall $1 million reduction in cost because the increase in productivity offsets the increased costs for training.[6]

---

[6]   A significant fraction of the cost savings does come from the force strength reduction of 150 personnel.

Constant Cost

Finally, we analyze the trade-off between productivity and force strength for a fixed cost. This analysis would likely be of most use internally to career field managers who want to understand training options given a set budget. In Figure 5.8, we hold total costs constant and allow productivity and work force size to vary. In this case, we see that a ten-day increase in course length leads to a productivity increase with a much smaller reduction in total force size. For example, at the same cost of $304 million, we can increase the course length by ten days, exactly offset the increased training costs by reducing the personnel strength by 15 airmen, and increase productivity from 3,700 to 3,850. In essence we have a tenfold increase in productivity—a gain of 150 in productivity for the loss of 15 personnel.

Conclusions

There are some caveats to this analysis. First, we do not take into account the need for integer cuts rather than fractional cuts in manpower authorizations. Proportional cuts in workforce size might sum to less than an integer in small work centers. A higher-resolution analysis would take into account this work-center size issue. Second, the results are heavily driven by the aggregate approximation we are using to represent changes in course length and the accompanying change in productivity. Last, the results depend on the validity of the productivity functions. Suggestions for improving our methodology to mitigate these concerns will be provided in the next chapter.

In this chapter, we used effectiveness curves and costs to develop a methodology for generating and selecting the preferred change (if any) to IST. We have shown how our methodology can be used to answer the following policy questions:

- For a given force size, what IST course length provides a work force at the least cost?
- Given a desired level of force effectiveness (productivity), how does IST course length impact the force size and cost?

- Given a fixed budget, what impact does IST course length have on the total force size and aggregate productivity?

This methodology can be used by policymakers who want to get the greatest "bang for the buck" for a fixed force strength, to trade cost and force strength while maintaining mission effectiveness, and to set course lengths for a fixed budget to achieve productivity and force strength goals. We performed this analysis for all seven specialties we analyzed. The corresponding plots are found in Appendix H.

CHAPTER SIX

Recommendations and Conclusions

This monograph has addressed the policy problem of how much enlisted training should be included in initial skills training and how much in on-the-job training. To make this distinction, we introduced the notion of the full cost of human capital development. The cost of IST is relatively easy to calculate because these training system costs are identifiable within the Air Force budget and because airmen-in-training are 100-percent dedicated to their training activities. The cost of OJT is much more difficult to capture—particularly the cost of an airman's time spent in training rather than in performing productive work. Using a survey instrument of career field experts, we developed productivity curves for this purpose. The survey instrument was also used to derive functions relating the change in IST course length to the increase or decrease in productivity at the end of IST. Finally, we combined these analyses to determine the impact of changes in the IST course length on cost, productivity, and end strength.

The results of this analysis have led us to the following major recommendations:

- *A significant increase in productivity for a small addition in IST course length is likely to exist for five of the seven specialties we analyzed.* In these cases, a ten-day increase in course length resulted in an improvement in the cost per unit of productivity. The largest gains were seen in AFSCs 3E7X1 and 3P0X1. Table 6.1 summarizes our results. For the example shown in Chapter Five, we

Table 6.1
Expanded Analysis Summary

AFSC	Specialty Title	Write-In Comments	Add vs. Drop Comparison	Incremental Change Functions	Steady State Analysis
1A8X1	Airborne Cryptologic Linguist	Don't add	Drop	Decrease	
1N3XX	Cryptologic Linguist		Drop	Decrease	
2T3X1	Special Purpose Vehicle Maintenance			Increase	Increase
2A3X3A	F-15 Tactical Aircraft Maintenance			Increase	Increase
2A5X1E	B-1/B-2 Aerospace Maintenance	Don't drop	Add	Increase	Increase
3P0X1	Security Forces	Don't drop	Add	Increase	Increase
3E7X1	Fire Protection		Add	Increase Don't decrease	Increase

estimated that an increase in IST course length for the 2T3X1 career field could decrease cost per unit of productivity by approximately $3,500 and overall HCD cost by $12 million. Significantly, the extra productivity of the early career force can reduce the aggregate force size requirement. For the two remaining specialties—AFSCs 1A8X1 and 1N3XX—neither increases nor decreases in course length resulted in a statistically significant reduction in the cost per unit of productivity. We believe there is evidence from the qualitative responses for considering reductions in IST for these specialties. Of particular interest are some of the specific changes suggested by survey respondents for additions to or removals from IST (see Appendix G).

- *Although we believe these increases are plausible the way we analyzed them, it would be prudent to replicate them with more specific and refined survey estimates.* As demonstrated in Chapter Five, our approach provides a means for determining the length of IST that provides the greatest cost-benefit ratio. This approach is dependent on the accuracy of the productivity curves associated with incremental changes in IST content and length. A preferred method for determining the incremental change function would be to calculate the change in productivity for a list of specific skills to be added to or deleted from IST. We also recommend that future productivity surveys be constructed with a top ten list of additions and a top ten list of deletions be provided to SMEs. We believe this would significantly improve estimates of the incremental change function.

- *The Air Force should also investigate other external measures to validate these productivity functions.* As a first step, a more uniform set of standards for defining productivity should be developed. Additionally, besides Operational Readiness Inspections (ORIs) or inspector general inspections, the Air Force could collect real production data on airmen of varying experience levels. The Air Force uses Organizational Management System (OMS) surveys to measure certain aspects of job/task knowledge/performance. In our evaluation of the OSR data, we did not find it sufficient to develop productivity curves. We recommend that AETC investigate building on current OMS tools to develop productivity curves and the functional relationship between incremental changes in IST content, course length, and graduate effectiveness.

- *This analysis also demonstrates the very large role played by the cost of OJT.* In many instances, OJT imposes very substantial costs that are not now factored into training policy decisions. Changes in IST course length can, in many cases, avoid an appreciable portion of those costs. *An important implication is that future assessments of course-length adjustments should explicitly consider effects on the extent and cost of OJT.* If they do not, the Air Force will be getting an incomplete and misleading picture of the consequences of policy. And the sums, as we have shown, can be large.

- *We recommend that the specific results of this analysis should be briefed to the respective AFSC U&TWs as they consider the suitability of the current IST curricula.* Although the cost-benefit analysis has produced only tentative indicators of possible payoff, we believe there is sufficient motivation for a closer examination of the course material that could be added or removed and the impact on initial productivity of training graduates, their subsequent learning curves, and total HCD cost.

- *Finally, we recommend that the AETC Studies and Analysis Squadron adopt the models and methodology developed in this study for future analyses involving the length of IST.* Other AFSCs can be easily examined to discover opportunities for reducing the total cost of training.

Air Force Specialty Code Descriptions

The following descriptions, as well as additional details about each specialty, can be found in Air Force Manual 36-2108, *Enlisted Classification* (Secretary of the Air Force, 1994a).

1A8X1 (Airborne Cryptologic Linguist): "Operates, evaluates, and manages airborne signals intelligence information systems and operations activities. Performs and supervises identification, acquisition, recording, translating, analyzing, and reporting of assigned voice communications. Provides signals intelligence threat warning support and interfaces with other units. Performs and assists in mission planning. Maintains publications and currency items. Maintains and supervises communication nets. Transcribes, processes, and conducts follow-up analysis of assigned communications." (p. 123)

1N3X1 (Cryptologic Linguist): "Performs and supervises acquisition, recording, transcribing, translating, analyzing, and reporting of assigned voice communications." (p. 155)

2A3X3A (F-15 Tactical Aircraft Maintenance): "Maintains tactical aircraft, support equipment, and forms and records. Performs and supervises flight chief, expediter, crew chief, repair and reclamation, quality assurance, and maintenance support functions." (p. 194)

2A5X1E (B-1/B-2 Aircraft Maintenance): "Maintains aircraft, support equipment (SE), and forms and records. Performs production supervisor, flight chief, expediter, crew chief, support, aero repair, and maintenance functions." (p. 194)

2T3X1 (Vehicle Maintenance): "Performs vehicle maintenance activities on military and commercial design general and special purpose, base maintenance, aircraft and equipment towing vehicles, and vehicular equipment. Activities include inspection, diagnostics, repair, and rebuild of components and assemblies." (p. 300)

3E7X1 (Fire Protection): "Protects people, property, and the environment from fires and disasters. Provides fire prevention, fire fighting, rescue, and hazardous material responses." (p. 367)

3P0X1 (Security Forces): "Leads, manages, supervises, and performs security force (SF) activities, including installation, weapon system, and resource security; antiterrorism; law enforcement and investigations; military working dog function; air base defense; armament and equipment; training; pass and registration; information security; and combat arms." (p. 395)

Survey Instrument

The figures that follow are screen shots of the actual survey used in the analysis. For the 1A8 specialty, some of the wording is different because the specialty uses aircrew terminology as opposed to the more common enlisted technical training terminology. Figure B.1 shows the questions used to determine the specialty background of the respondent. The pull-down menu in question 1 allowed the respondent to select one of the seven specialties investigated in this analysis: 1A8X1, 1N3XX, 2A3X3A, 2A5X1E, 2T3X1, 3E7X1, and 3P0X1.

Figure B.2 contains the core concept and question of the survey. The definition of fully mission-effective worker is a critical reference for this and the successive questions. The pull-down menus referring to rank allowed the user to select from E-1 to E-9. Minimum and maximum responses allowed the individual to specify a range rather than just a single, average number. In pre-tests of the survey, we found than many SMEs were more comfortable specifying a range rather than a single number.

Figure B.1
Survey Questions 1–3

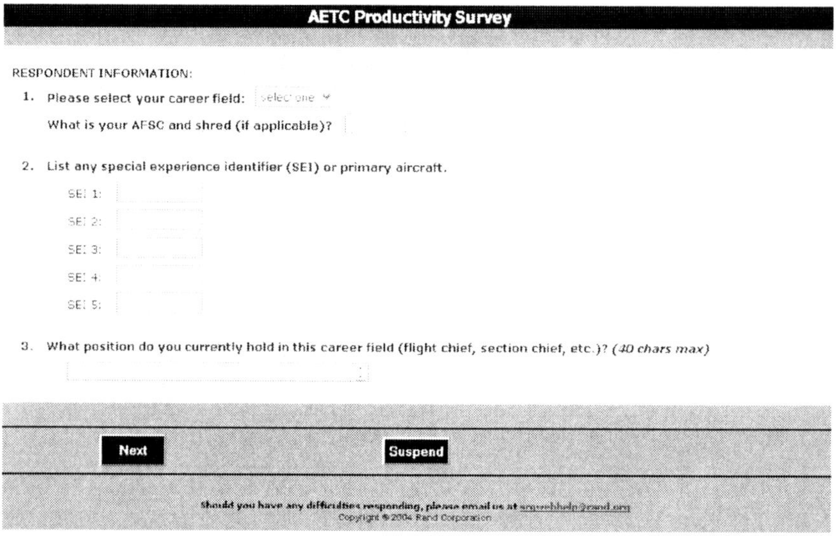

AETC Productivity Survey

RESPONDENT INFORMATION:

1. Please select your career field: [select one ▼]

   What is your AFSC and shred (if applicable)?

2. List any special experience identifier (SEI) or primary aircraft.

   SEI 1:
   SEI 2:
   SEI 3:
   SEI 4:
   SEI 5:

3. What position do you currently hold in this career field (flight chief, section chief, etc.)? (*40 chars max*)

[Next]   [Suspend]

Should you have any difficulties responding, please email us at arqwebhelp@rand.org
Copyright © 2004 Rand Corporation

RAND *MG555-B.1*

Figure B.2
Survey Question 4

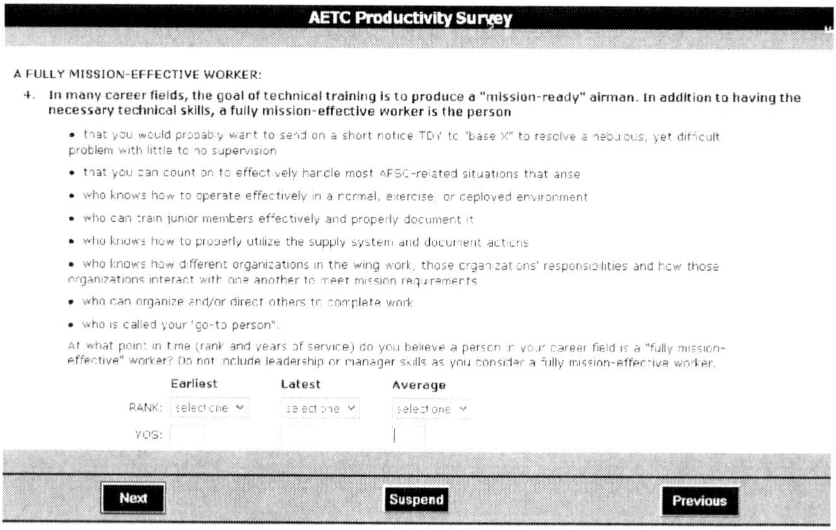

AETC Productivity Survey

A FULLY MISSION-EFFECTIVE WORKER:

4. In many career fields, the goal of technical training is to produce a "mission-ready" airman. In addition to having the necessary technical skills, a fully mission-effective worker is the person

   • that you would probably want to send on a short notice TDY to "base X" to resolve a nebulous, yet difficult problem with little to no supervision
   • that you can count on to effectively handle most AFSC-related situations that arise
   • who knows how to operate effectively in a normal, exercise, or deployed environment
   • who can train junior members effectively and properly document it
   • who knows how to properly utilize the supply system and document actions
   • who knows how different organizations in the wing work, those organizations' responsibilities and how those organizations interact with one another to meet mission requirements
   • who can organize and/or direct others to complete work
   • who is called your 'go-to person'.

   At what point in time (rank and years of service) do you believe a person in your career field is a "fully mission-effective" worker? Do not include leadership or manager skills as you consider a fully mission-effective worker.

	Earliest	Latest	Average
RANK:	select one ▼	select one ▼	select one ▼
YOS:			

[Next]   [Suspend]   [Previous]

RAND *MG555-B.2*

Figure B.3 shows a free-response box that allowed the respondent to recommend alternative views on the definition of a fully mission-effective worker. It also contains the questions used to determine the shape of the productivity curves. Originally, we had hoped to provide respondents with an interactive graph that they could manipulate in order to draw their own form of the productivity curve. Doing so proved to be too difficult, so this alternative method was selected.

Figure B.4 shows the questions asked in a different way, to gain insight into the form of the productivity curve. Instead of asking for a specific YOS, we asked the respondent to determine when level certifications occur, and what productivity is achieved when those certification levels are completed. That way, we could infer additional data points for YOS and productivity.

We realized that our definition of fully mission-effective worker did not include skills specific to leadership and management. Those skills are acquired at the same time as specific specialty skills are also

Figure B.3
Survey Questions 5–6

AETC Productivity Survey		

5. Do you believe the description of being "fully mission-effective" is accurate? If not, what is lacking or overstated? How would you define "fully mission-effective"? *(1000 chars max)*

6. In question 4, you said that it takes between 4 and 4 years to be a fully mission-effective worker. As an airman progresses through each the years of service listed below, how effective are they on average relative to the "fully mission-effective" or 100% person?
    Example: suppose the airmen you have observed with 0–1 YOS are between 10% and 25% effective with an average of 20% effective with regard to the fully effective person in question 4. On line a., you would type in 10% under Least Effective Person, 25% under Most Effective Person, and 20% under Average Person.

	Least Effective Person	Most Effective Person	Average Person
a. End of 1 YOS?	%	%	%
b. End of 3 YOS?	%	%	%
c. End of 5 YOS?	%	%	%
d. End of 8 YOS?	%	%	%
e. End of 12 YOS?	%	%	%

Figure B.4
Survey Questions 7–8

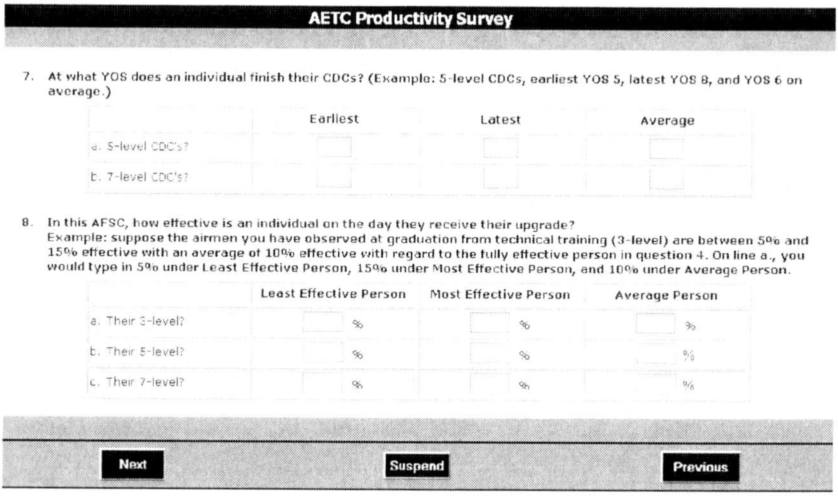

7. At what YOS does an individual finish their CDCs? (Example: 5-level CDCs, earliest YOS 5, latest YOS 8, and YOS 6 on average.)

	Earliest	Latest	Average
a. 5-level CDC's?			
b. 7-level CDC's?			

8. In this AFSC, how effective is an individual on the day they receive their upgrade?
Example: suppose the airmen you have observed at graduation from technical training (3-level) are between 5% and 15% effective with an average of 10% effective with regard to the fully effective person in question 4. On line a., you would type in 5% under Least Effective Person, 15% under Most Effective Person, and 10% under Average Person.

	Least Effective Person	Most Effective Person	Average Person
a. Their 3-level?	%	%	%
b. Their 5-level?	%	%	%
c. Their 7-level?	%	%	%

Next   Suspend   Previous

RAND MG555-B.4

being developed. Therefore in Figure B.5, we asked questions about the rank and YOS needed to become a fully mission-effective leader/ manager. To determine the increase over time in level of managerial effectiveness, we also asked about effectiveness for different ranks.

OJT effectiveness is intrinsically linked with the availability of supervisors to provide training and oversight. Questions 11 and 12 in Figure B.6 assess the amount of time supervisors spend in training and the rank and YOS of those supervising OJT. We realized that OJT supervisors are sometimes selected from lower grades or with fewer YOS because of manpower shortages or high operations tempo. Therefore, we asked what rank and YOS the ideal supervisor should have.

Figure B.5
Survey Questions 9–10

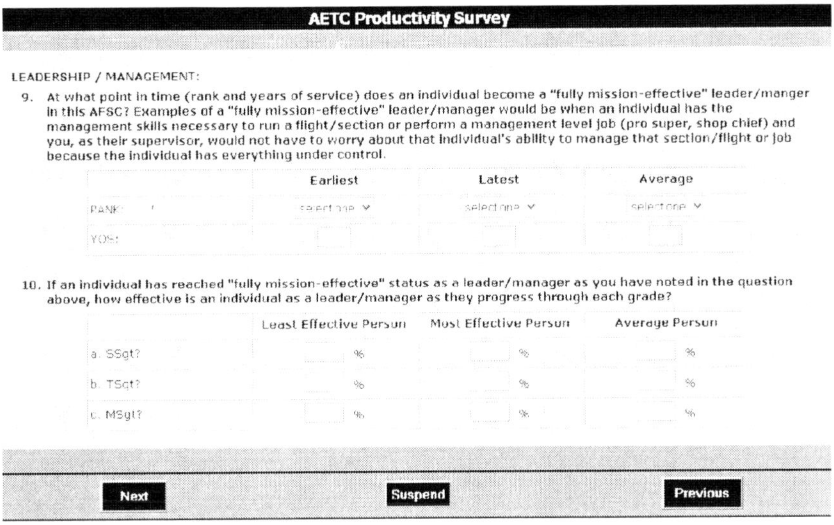

AETC Productivity Survey

LEADERSHIP / MANAGEMENT:

9. At what point in time (rank and years of service) does an individual become a "fully mission-effective" leader/manger in this AFSC? Examples of a "fully mission-effective" leader/manager would be when an individual has the management skills necessary to run a flight/section or perform a management level job (pro super, shop chief) and you, as their supervisor, would not have to worry about that individual's ability to manage that section/flight or job because the individual has everything under control.

	Earliest	Latest	Average
RANK:	select one ∨	select one ∨	select one ∨
YOS:			

10. If an individual has reached "fully mission-effective" status as a leader/manager as you have noted in the question above, how effective is an individual as a leader/manager as they progress through each grade?

	Least Effective Person	Most Effective Person	Average Person
a. SSgt?	%	%	%
b. TSgt?	%	%	%
c. MSgt?	%	%	%

Next	Suspend	Previous

RAND MG555-B.5

Figure B.6
Survey Questions 11–12

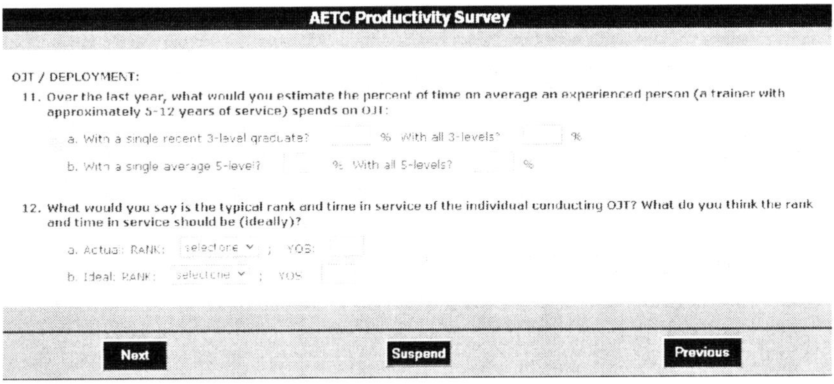

AETC Productivity Survey

OJT / DEPLOYMENT:

11. Over the last year, what would you estimate the percent of time on average an experienced person (a trainer with approximately 5-12 years of service) spends on OJT:

    a. With a single recent 3-level graduate? ___% With all 3-levels? ___%

    b. With a single average 5-level? ___% With all 5-levels? ___%

12. What would you say is the typical rank and time in service of the individual conducting OJT? What do you think the rank and time in service should be (ideally)?

    a. Actual: RANK: select one ∨ ; YOS: ___

    b. Ideal: RANK: select one ∨ ; YOS: ___

Next	Suspend	Previous

RAND MG555-B.6

The number of trainees in a unit, as well as the overall unit manning, affect both supervisors and trainees. Figure B.7 and Figure B.8 address these issues by asking for reductions in productivity for the trainer and the trainee given increasing numbers of trainees and a decreasing percentage of manning.

Deployments also have an effect on the ability of an airman to increase productivity. Question 17 in Figure B.9 asks how training is affected when different combinations of trainee and trainer deployments occur. Furthermore, we asked in question 18 how effective an airman needs to be in order to successfully deploy.

Figure B.7
Survey Questions 13–14

REDUCED PRODUCTIVITY OF TRAINERS

For the next two questions, we define the cost of OJT as the amount of the trainer's lost effectiveness or lost productivity incurred when this experienced individual is either removed from productive work to conduct OJT or must simultaneously conduct OJT and perform regular duties. For example, if an experienced baker could make 10 pies on a normal day without OJT responsibilities, but only 8 pies when conducting a full day of OJT with a single trainee, the reduction in his productivity when conducting OJT is 20%.

13. What is the reduced productivity of the trainer with the following number of trainees: (0% would mean no effect, 10% would mean 10% less effective in conducting OJT, etc. Please indicate DNA if the number of trainee does not occur.)

	Reduced Productivity	
1 trainee:	%	
2 trainees:	%	☐ DNA
3 trainees:	%	☐ DNA
4 trainees:	%	☐ DNA
5 trainees:	%	☐ DNA

14. Assuming no surging, overtime, or additional hours, what is the reduced productivity of the trainer when 5/7 level manning falls below 100% (at your duty station) for each of the various scenarios? (0% would mean no effect, 10% would mean 10% less effective in conducting OJT, etc.)

	Reduced Productivity
a. 90% manning in 5/7 levels:	%
b. 80% manning in 5/7 levels:	%
c. 70% manning in 5/7 levels:	%
d. 60% manning in 5/7 levels:	%
e. 50% manning in 5/7 levels:	%

RAND MG555-B.7

Figure B.8
Survey Questions 15–16

REDUCED QUALITY OF OJT

For the next 2 questions we would like to better understand the impact on the quality of OJT when either a heavier trainee-to-trainer load is incurred or manning is reduced. By quality, we mean the ability to teach mission-relevant skills in both a timely and a thorough manner. We assume that one-on-one training at 100% manning provides the baseline for the best OJT quality that can be achieved. Please answer the next two questions with this baseline in mind.

15. What is the reduced quality of the OJT experience while a trainer performs OJT with the following number of trainees:
(Assume 0% reduction with 1 trainee for each trainer. 0% would mean no effect, 10% would mean 10% less effective in conducting OJT, etc. Please indicate DNA if the number of trainee does not occur.)

	Reduced OJT Quality	
2 trainees:	%	☐ DNA
3 trainees:	%	☐ DNA
4 trainees:	%	☐ DNA
5 trainees:	%	☐ DNA

16. Assuming no surging, overtime, or additional hours, what is the reduced quality of OJT when 5/7 level manning falls below 100% (at your duty station) for each of the various scenarios? (0% would mean no effect, 10% would mean 10% less effective in conducting OJT, etc.)

	Reduced OJT Quality
f. 90% manning in 5/7 levels:	%
g. 80% manning in 5/7 levels:	%
h. 70% manning in 5/7 levels:	%
i. 60% manning in 5/7 levels:	%
j. 50% manning in 5/7 levels:	%

RAND *MG555-B.8*

Figure B.9
Survey Questions 17–18

AETC Productivity Survey

17. If your unit deploys, what is the impact on 3-level OJT effectiveness? (0% would mean no effect, 10% would mean 10% less effective in conducting OJT, etc.)

a. Trainee deploys; trainer deploys: %   ☐ DNA

b. Trainee deploys; trainer does not: %   ☐ DNA

c. Trainee does not deploy; trainer deploys: %   ☐ DNA

18. What is the required level of mission effectiveness needed before a 3-level airman should deploy? Remember that 100% refers to a fully mission-effective worker.   %

Next		Suspend		Previous

RAND *MG555-B.9*

The final questions, 19 and 20, shown in Figure B.10, provide the critical information on how to determine the effect on course graduates' productivity of adding or deleting course elements from IST. The questions ask for the amount of time required in both IST and OJT to learn this course element.

Figure B.10
Survey Questions 19–20

AETC Productivity Survey

TECHNICAL TRAINING:

19. If you had to add instruction or increase the length of technical training, what would you add? (Consider the entire 3-level pipeline training.)

   a. Provide a descriptive name for the group of instruction you would add to technical training

   b. How long would it take to teach this material at technical training (estimated in days)? _____ days

   c. What would be the new effectiveness of a technical training graduate with this instruction? (Recall that you said that the current trainee, on average, is 50% effective.) _____ %

   d. If it takes one day to earn the average task you would like to add, how many days (on average) would it take a person doing OJT to learn the same task or subject (given the availability of resources and all the normal demands and time constraints of a functioning unit)? _____ days

20. If you are forced to eliminate instruction or decrease the length of technical training, what would you eliminate? (Consider the entire 3-level pipeline training.)

   a. Provide a descriptive name for the group of instruction you would take out of technical training.

   b. How long does it currently take to teach this material in technical training? (estimated in days)? _____ days

   c. What would be the new effectiveness of a technical training graduate without this instruction? (Recall that you said that the current trainee, on average, is 50% effective.) _____ %

   d. If it takes one day to earn the average task you would like to eliminate, how many days (on average) would it take a person doing OJT to learn the same task or subject (given the availability of resources and all the normal demands and time constraints of a functioning unit)? _____ days

This is the last question of the survey. Once you hit "Next" all of your responses will be transferred and you will no longer have access to your responses. Please note that you must click "Next" in order for your survey to be considered completed.
THANK YOU FOR YOUR PARTICIPATION!

RAND MG555-B.10

Defining "Fully Mission-Effective" or "Mission-Ready" Airmen

After examining survey results, we realized that respondents reacted differently to specific terminology. For example, throughout our survey, we use the word "airman" to refer to an enlisted soldier of any rank. Unfortunately, this use of the word confused many respondents, who were using the word "airman" to refer to a soldier holding one of the junior enlisted ranks.

In this analysis, we relied on a comprehensive and consistent definition for "fully mission effective" and used the terms "fully mission effective" and "100-percent productive" interchangeably throughout the survey. After seeing the survey responses, we realized several shortcomings of this decision. First, we allowed respondents to comment on and revise our definition of fully mission effective. Not everyone took the opportunity to provide comments (on average, 37.7 percent of respondents left the field blank) but the responses of those who did merit discussion.

Overall, approximately 26 percent of respondents agreed that our definition was accurate. However, a comparable number in each specialty felt that we had either over- or understated what they would consider a fully mission-effective worker to be. In addition, there were individuals who felt that the definition of fully mission effective should be interpreted as job- and/or rank-dependent—once an individual reaches 100-percent productivity at one job, he/she gets promoted or moves to a new job and takes on a new set of responsibilities. Finally, there were individuals who adamantly said that there is no such thing as a 100-

percent productive worker. Table C.1 summarizes these responses for each specialty.

If we assume that the respondents who left the comment field blank were satisfied with the definition, more than half of the respondents in each specialty used the same working definition of 100-percent productive to respond to the survey questions.

The definition of fully mission effective was derived from prior research done in the aircraft maintenance specialty. Respondents in the nonmaintenance fields did make comments about the irrelevance of some parts of our definition for fully mission effective. Although we did modify the survey to make the terminology more appropriate for the 1A8X1 specialty, it is possible that a specialty-specific definition for 100-percent productive would have been more useful.

Lessons Learned in Defining "Fully Mission Effective"

One oversight, which we noticed and corrected early in the survey response period, was that we did not include a sizable comment block at the conclusion of the survey for individuals to voice concerns or feedback about the overall survey. Once we had introduced that feature into the survey, more than 60-percent took advantage of the opportunity to respond at length about survey issues. This comment block was particularly helpful for identifying respondents who lacked the proper experience for their survey data to be worthwhile. Respondents also used this block to voice other concerns about general training issues within their specialty.

An additional problem was that the open-ended sections of questions 19 and 20 were not long enough for all the respondents to completely describe the blocks of training that they would add or subtract. Although in most cases it was relatively easy to identify the CFETP training module that the respondent was describing, there were observations in which the description stopped in the middle of a word. While this did not affect any of the quantitative research in the body of the report, it does increase the uncertainty of the completeness of the comments described in Appendix G.

Table C.1
Responses to the Term "Fully Mission Effective" (%)

	1A8X1	1N3XX	3E7X1	3P0X1	2A5X1E	2A3X3A	2T3X1	Average
Accurate	29	34	21	24	25	24	27	26
Blank	41	21	34	41	44	39	43	38
Overstated	12	17	21	11	13	13	12	14
Understated	18	23	10	14	11	18	15	17
Job/rank dependent	0.0	1.4	3.4	4.5	3.2	4.4	1.4	2.6
No such thing	0.0	2.9	0.0	6.3	3.2	1.5	1.4	2.2

Data Cleaning

The survey used in this analysis reached a wide audience and included a wide range of responses. It was therefore necessary to carefully examine the response data for appropriateness. Our data cleaning process consisted of two separate steps. The first removed entire respondents from the data set. The second removed particular responses but retained as many of the respondents' individual answers as possible. The criteria for each of these kinds of removals are described in more detail below.

First, we removed from consideration any respondents who indicated that they were no longer active in the specialty from which they were invited to participate. These respondents had crossflowed into a different AFSC. They were identified by either their written comments or because they specified an AFSC in question 2 that did not match the AFSC selected from the drop-down menu in the first question. In addition, two respondents were deleted in their entirety based on their answers. One of these respondents answered "1" in every possible response blank. The other appeared to have answered the survey twice since the responses matched on a character-for-character basis throughout the entire survey. Overall, 70 of the original 865 respondents were eliminated from our analysis. Table D.1 shows the total number sampled, the number of initial respondents received and the number actually used in the analysis. Only 2–15 percent of the surveys received were dropped.

Table D.1
Response Rate After Initial Check

AFSC	Number Sampled	Initial Responses	Final Responses	Final Response Rate (%)
1A8X1	130	110	105	78
1N3XX	400	69	60	15
2T3X1	400	137	129	32
2A3X3A	400	135	125	31
2A5X1E	400	157	153	38
3P0X1	400	110	100	25
3E7X1	400	146	123	31

The second portion of the data cleaning process eliminated responses to specific questions for a specific respondent. We performed several checks to ensure that the survey responses were logical and accurate. The following chart illustrates the percentage of question blanks that were filled before and after the data cleaning process. When we began the data cleaning process, approximately 20 percent of the questions were unanswered. After the data cleaning procedures, approximately 25–40 percent (Figure D.1) of the response fields had missing data. The data cleaning processes for each question are described below.

For questions 6, 7, and 8, which asked for a minimum, average, and maximum productivity at several time points, the following checks were performed:

1. Determine if minimum ≤ average ≤ maximum for all time points.
2. For each minimum, average, or maximum, determine if effectiveness erroneously decreased over time.
3. Determine if the difference in maximum and minimum productivity exceeded 99 percent. In such cases, we interpreted this to mean that the respondent had no idea what the productivity should be.
4. Determine if the difference in maximum and minimum YOS exceeds 19 years. This, too, suggests that the respondent had no idea how many years should be specified.

Figure D.1
Data Cleaning Results

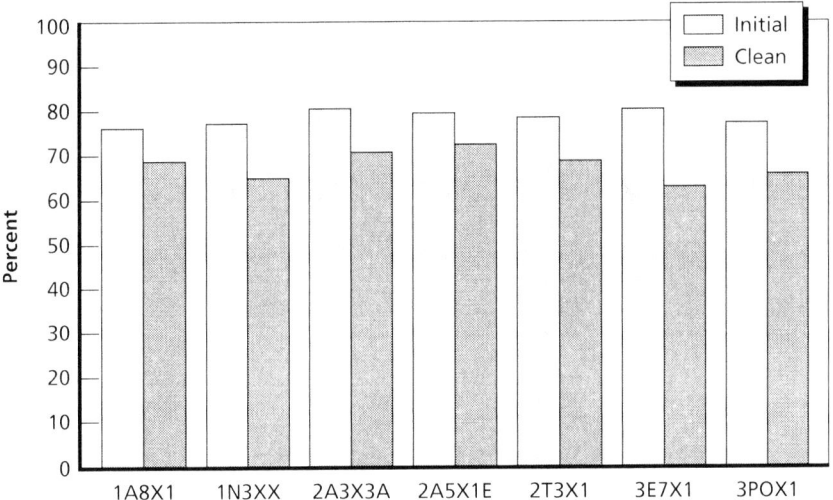

NOTE: 1A8X1 = Airborne Cryptologic Linguist; 1N3XX = Cryptologic Linguist;
2A3X3A = F-15 Maintenance; 2A5X1E = B-1/B-2 Maintenance; 2T3X1 = Special
Purpose Vehicle Maintenance; 3E7X1 = Fire Protection; 3POX1 = Security Forces.
RAND *MG555-D.1*

In cases where these criteria were not met, the erroneous values in the sequence were set to missing. In particular, we felt that numbers 3 and 4 were necessary because including minimum and maximum observations with such wide ranges provided no useful information for our analysis and functioned only to increase the standard deviation of the values that were included. By setting these responses to missing, we are able to make more accurate inferences regarding the data.

Question 6 requested productivity levels at increasing YOS increments. For this question, we excluded observations indicating that the maximum productivity of an airman with 8 or 12 YOS was less than 50 percent or that the airman's productivity never exceeded 10 percent.

Question 7 required respondents to indicate the number of years necessary to acquire a 5- and 7-level skill upgrade. For this question, we excluded observations that indicated a time to 5-level greater than 12 years and excluded observations that indicated a time to 7-level less than four years.

Question 8 requested the minimum, average, and maximum productivity on award of 3-, 5- and 7-level. Here again, we eliminated responses indicating that the productivity never exceeded 10 percent.

In question 10, we asked how effective the average airman is as a leader or manager at various grades. We eliminated any observations that had a value of 20 or less for each of the ranks, because those answers seem to indicate a misunderstanding of the question being posed. Rather than answer in terms of productivity, it appeared that these respondents answered in terms of years to achieve a certain rank. We also excluded observations where there was no variation in the minimum, maximum, or average across all the ranks. For the purposes of our analysis, we would anticipate at least some growth in leadership productivity as individuals progress from E-6 to E-9.

In questions 13 and 14, we asked respondents to estimate the change in the productivity of the trainer and quality of the OJT for the trainee as the number of trainees per trainer increases. Although we recognize the possibility that productivity and OJT quality could remain the same regardless of the number of trainees per trainer, we do not believe that either productivity or quality would increase with an increase in the ratio of students to trainers. Therefore, we assumed that any observations indicating an increase in productivity or OJT quality as that ratio increased were based on a misunderstanding of the question and were set to missing. Similarly, in questions 15 and 16, observations that indicated an increase in OJT quality or trainer productivity as the trainer manning level decreased were also set to missing.

Statistical Results

In this appendix we provide three tables of statistical data related to our productivity curves. Table E.1 presents descriptive statistics on the central tendency and spread (standard deviation and inter-quartile range) of the survey responses. Included for each AFSC are the years specified in the survey (1, 3, 5, 8, and 12) along with three other years in decimal form (two for 1A8) representing graduation from IST, 5-level advancement, and 7-level advancement. These data can be used to replicate the productivity curves used in the cost-benefit analysis. Table E.2 shows the predicted values after curve fitting (the values on the left are the original values predicted by the curve fitting; those on the right have been transformed back onto a 0–100 scale). Table E.3 shows the estimated values (along with standard errors and other statistics) for the five parameters required by the curve-fitting function.

Table E.1
Statistical Measures of Effectiveness Data

AFSC	Year	Mean	Std. Dev.	Std. Error	IQR
1A8X1	1	9.947	11.280	1.163	15
	1.392	25.729	19.019	1.941	27.5
	3	49.591	20.847	2.162	37
	4.134	52.321	21.256	2.362	28
	5	73.968	16.121	1.672	20
	8	86.033	12.558	1.316	15
	12	91.180	10.272	1.089	15
1N3XX	1	11.618	13.467	1.816	20
	1.019	23.825	19.428	2.573	20
	3	48.818	21.708	2.927	20
	3.492	56.571	18.229	2.436	25
	5	71.509	19.402	2.616	28
	6.373	78.179	15.454	2.065	20
	8	83.382	14.964	2.018	20
	12	89.873	10.358	1.397	14
2T3X1	0.438	20.858	13.991	1.277	20
	1	25.947	14.961	1.401	15
	2.387	61.149	15.593	1.460	25
	3	56.486	20.356	1.932	30
	5	76.028	18.043	1.736	20
	6.806	78.965	11.980	1.127	15
	8	86.981	13.736	1.340	17
	12	92.160	10.135	0.984	10
2A3X3A	0.554	21.198	15.422	1.432	17.5
	1	24.383	15.794	1.473	15
	2.254	54.045	17.314	1.643	25
	3	56.250	20.499	1.937	30
	5	74.536	18.582	1.756	25
	5.948	80.009	14.331	1.348	15
	8	85.620	15.661	1.507	15
	12	92.383	11.623	1.124	10
2A5X1E	0.346	21.000	15.380	1.286	15
	1	26.752	14.988	1.281	15
	1.875	59.860	18.574	1.593	27.5
	3	64.612	21.097	1.823	35
	5	80.962	16.648	1.460	23
	5.811	83.000	14.014	1.206	20
	8	88.444	13.964	1.244	15
	12	93.375	11.130	1.016	10
3P0X1	0.323	22.611	15.036	1.585	20
	1	23.306	16.139	1.751	10
	2.117	51.989	19.718	2.102	25
	3	48.536	20.393	2.225	31.5
	5	65.241	20.257	2.223	30
	5.917	74.841	14.055	1.498	20
	8	77.100	17.080	1.910	20
	12	87.139	12.882	1.449	15

Table E.1—continued

AFSC	Year	Mean	Std. Dev.	Std. Error	IQR
3E7X1	0.377	25.478	17.065	1.605	25
	1	23.785	14.958	1.446	15
	2.085	52.303	19.165	1.926	25
	3	50.077	19.689	1.931	30
	5	67.192	20.083	2.018	25
	6.291	75.857	15.194	1.535	15
	8	79.031	17.256	1.761	15
	12	88.579	13.625	1.398	14

Table E.2
Predicted/Back-Transformed Values and Confidence Limits

AFSC	Year	Predicted Value			Back-Transformed Predicted Value		
			Confidence Limit			Confidence Limit	
		Base	Lower 95%	Upper 95%	Base	Lower 95%	Upper 95%
1A8X1	1	0.363	0.336	0.390	12.588	10.853	14.433
	1.392[a]	0.425	0.398	0.451	16.984	15.048	19.012
	3	0.717	0.688	0.746	43.168	40.297	46.063
	4.134[b]	0.914	0.879	0.949	62.726	59.286	66.104
	5	1.035	0.998	1.073	73.957	70.580	77.197
	8	1.244	1.202	1.286	89.719	87.038	92.122
	12	1.295	1.247	1.343	92.584	89.862	94.911
1N3XX	1	0.393	0.358	0.428	14.690	12.297	17.257
	1.019[a]	0.396	0.361	0.431	14.911	12.504	17.491
	3	0.760	0.717	0.803	47.445	43.139	51.771
	3.492[b]	0.846	0.798	0.893	56.021	51.287	60.700
	5	1.050	0.996	1.105	75.270	70.414	79.824
	6.373[c]	1.156	1.098	1.213	83.745	79.303	87.744
	8	1.215	1.153	1.276	87.842	83.559	91.555
	12	1.249	1.182	1.316	89.997	85.619	93.654
2T3X1	0.438[a]	0.481	0.455	0.507	21.407	19.349	23.539
	1	0.570	0.547	0.593	29.085	27.026	31.188
	2.387[b]	0.794	0.767	0.820	50.820	48.173	53.464
	3	0.885	0.855	0.915	59.866	56.892	62.805
	5	1.107	1.073	1.141	79.987	77.186	82.648
	6.806[c]	1.211	1.177	1.245	87.598	85.269	89.752
	8	1.246	1.211	1.282	89.840	87.582	91.894
	12	1.286	1.244	1.327	92.092	89.698	94.193

Table E.2—continued

AFSC	Year	Predicted Value			Back-Transformed Predicted Value		
			Confidence Limit			Confidence Limit	
		Base	Lower 95%	Upper 95%	Base	Lower 95%	Upper 95%
2A3X3A	0.554[a]	0.477	0.454	0.501	21.116	19.196	23.102
	1	0.545	0.522	0.568	26.894	24.889	28.948
	2.254[b]	0.744	0.719	0.770	45.908	43.409	48.416
	3	0.858	0.829	0.887	57.240	54.364	60.091
	5	1.096	1.061	1.131	79.112	76.221	81.863
	5.948[c]	1.169	1.134	1.204	84.710	82.078	87.168
	8	1.259	1.221	1.297	90.614	88.278	92.715
	12	1.310	1.266	1.354	93.361	90.999	95.385
2A5X1E	0.346[a]	0.471	0.447	0.495	20.581	18.669	22.562
	1	0.599	0.577	0.621	31.821	29.797	33.879
	1.875[b]	0.777	0.751	0.802	49.135	46.585	51.687
	3	0.979	0.946	1.012	68.845	65.756	71.853
	5	1.198	1.163	1.233	86.723	84.283	88.987
	5.811[c]	1.243	1.208	1.278	89.610	87.384	91.643
	8	1.298	1.260	1.336	92.726	90.634	94.571
	12	1.316	1.275	1.357	93.658	91.513	95.509
3P0X1	0.323[a]	0.487	0.457	0.516	21.865	19.486	24.342
	1	0.567	0.539	0.595	28.849	26.343	31.421
	2.117[b]	0.703	0.673	0.733	41.798	38.841	44.785
	3	0.806	0.772	0.841	52.099	48.665	55.524
	5	1.001	0.959	1.042	70.873	67.040	74.562
	5.917[c]	1.066	1.023	1.109	76.634	72.914	80.158
	8	1.165	1.119	1.212	84.436	80.940	87.637
	12	1.238	1.182	1.294	89.328	85.649	92.519
3E7X1	0.377[a]	0.498	0.471	0.524	22.791	20.632	25.026
	1	0.570	0.545	0.595	29.146	26.893	31.452
	2.085[b]	0.700	0.673	0.727	41.515	38.893	44.162
	3	0.807	0.776	0.837	52.111	49.048	55.166
	5	1.003	0.966	1.041	71.114	67.672	74.438
	6.291[c]	1.095	1.056	1.135	79.057	75.791	82.145
	8	1.178	1.136	1.219	85.330	82.282	88.136
	12	1.260	1.210	1.310	90.640	87.516	93.355

[a] 3-level; [b] 5-level; [c] 7-level.

Table E.3
Parameter Estimates

AFSC	Parameter	Estimate	Std. Error	Confidence Limit		t-Value	p-value
				Lower 95%	Upper 95%		
1A8X1	B_1	1.301	0.025	1.251	1.351	51.834	<0.001
	B_2	4.609	0.299	4.016	5.202	15.433	<0.001
	B_3	0.578	0.026	0.525	0.630	21.963	<0.001
	σ_u^2	0.031	0.006	0.019	0.044	5.155	<0.001
	σ_ε^2	0.033	0.002	0.029	0.037	16.333	<0.001
1N3XX	B_1	1.252	0.034	1.184	1.321	36.726	<0.001
	B_2	4.009	0.306	3.395	4.623	13.080	<0.001
	B_3	0.607	0.035	0.537	0.678	17.295	<0.001
	σ_u^2	0.047	0.010	0.027	0.068	4.702	<0.001
	σ_ε^2	0.028	0.002	0.024	0.032	13.882	<0.001
2T3X1	B_1	1.292	0.022	1.248	1.336	58.396	<0.001
	B_2	2.105	0.085	1.937	2.272	24.893	<0.001
	B_3	0.507	0.028	0.452	0.561	18.364	<0.001
	σ_u^2	0.032	0.005	0.022	0.042	6.405	<0.001
	σ_ε^2	0.024	0.001	0.022	0.026	19.531	<0.001
2A3X3A	B_1	1.319	0.023	1.273	1.366	56.232	<0.001
	B_2	2.307	0.086	2.136	2.478	26.735	<0.001
	B_3	0.486	0.022	0.442	0.529	21.981	<0.001
	σ_u^2	0.038	0.006	0.026	0.050	6.451	<0.001
	σ_ε^2	0.024	0.001	0.022	0.026	19.536	<0.001

Table E.3—continued

AFSC	Parameter	Estimate	Std. Error	Confidence Limit Lower 95%	Upper 95%	t-Value	p-value
2A5X1E	B_1	1.318	0.021	1.276	1.360	62.676	<0.001
	B_2	2.230	0.084	2.064	2.395	26.657	<0.001
	B_3	0.620	0.029	0.563	0.678	21.207	<0.001
	σ_u^2	0.043	0.006	0.031	0.055	7.278	<0.001
	σ_ε^2	0.024	0.001	0.022	0.027	21.269	<0.001
3P0X1	B_1	1.259	0.032	1.196	1.322	39.843	<0.001
	B_2	1.800	0.073	1.655	1.945	24.691	<0.001
	B_3	0.388	0.024	0.341	0.436	16.341	<0.001
	σ_u^2	0.048	0.008	0.031	0.064	5.664	<0.001
	σ_ε^2	0.024	0.001	0.021	0.027	16.984	<0.001
3E7X1	B_1	1.286	0.029	1.228	1.344	44.064	<0.001
	B_2	1.823	0.065	1.695	1.951	28.198	<0.001
	B_3	0.374	0.021	0.333	0.415	18.020	<0.001
	σ_u^2	0.048	0.008	0.033	0.063	6.278	<0.001
	σ_ε^2	0.024	0.001	0.021	0.026	18.458	<0.001

Productivity Curves

In this appendix, we provide a detailed explanation of how we developed the specialty productivity curves.

Methodology for Fitting Functions to the Productivity Data

When fitting curves based on percentages with a significant number of responses outside the range of 30 percent to 70 percent, it is common practice to normalize the data using an arcsine transformation (Sokal and Rohlf, 1995). The more observations outside this range (closer to 0 percent or 100 percent), the more normality is violated and the stronger the recommendation to use the arcsine transformation. However, the arcsine transformation is not effective when a substantial number of observations are at 0 percent or 100 percent or when the sample size is small. In our case, higher years of experience will have a substantial number of 100 percent responses, so the arcsine square root transformation is warranted (Sokal and Rohlf, 1995). Thus, to make proper statistical inferences on our data, each effectiveness, E, is transformed to y:

$$y = \arcsin \sqrt{E} \, .$$

By performing this transformation, the variability of the data becomes rather small and a better fit is achieved for each specialty/year combination, yielding more normally distributed residuals.

We explored several different methods for fitting the transformed data. One method fit a nonlinear mixed model. Since an individual's responses are dependent (i.e., how they respond to a subsequent year depends on how they responded the previous year), this approach seemed ideal. By plotting effectiveness over time, one can see a mild nonlinear trend along with a marked increase in variability over time. The following logistic growth model takes into account the dependence between an individual's responses. This model was fit for each specialty:

$$y_{ij} = \frac{B_1 + u_i}{1 + B_2 * e^{(-B_3 * x_{ij})}} + \varepsilon_{ij}$$

where

$y_{ij} = \arcsin \sqrt{E_{ij}}$

y_{ij} = the observation for the jth year for the ith subject

x_{ij} = the jth year for the ith subject

B_1, B_2, and B_3 = the fixed-effect parameters that minimize the squared residual random error ε_{ij}.

B_1 estimates the upper horizontal asymptote.

$\dfrac{B_1}{1 + B_2}$ estimates the y-intercept.

B_3 determines the shape of the curve.

u_i, the random coefficient parameter, models the subject-to-subject random variation of the upper horizontal asymptote. In other words, it captures the effectiveness that a typical subject in a given specialty approaches over time. This parameter is assumed to follow a normal distribution with mean 0 and variance σ_u^2. Note that ε_{ij} are the residual random errors and are assumed i.i.d N $(0, \sigma_\varepsilon^2)$.

For each specialty, the logistic growth model was fit with different starting estimates for B_1, B_2, B_3, σ_u^2, and σ_ε^2. Since we sampled only at specific years, we calculated values for the rest of the years to get a full set of predicted values. Once we obtained our predicted values and estimated a 95-percent confidence interval (CI) of the mean, the inverse

Figure F.1
Plot of Back-Transformed Predicted Values and CIs, AFSC 2T3X1

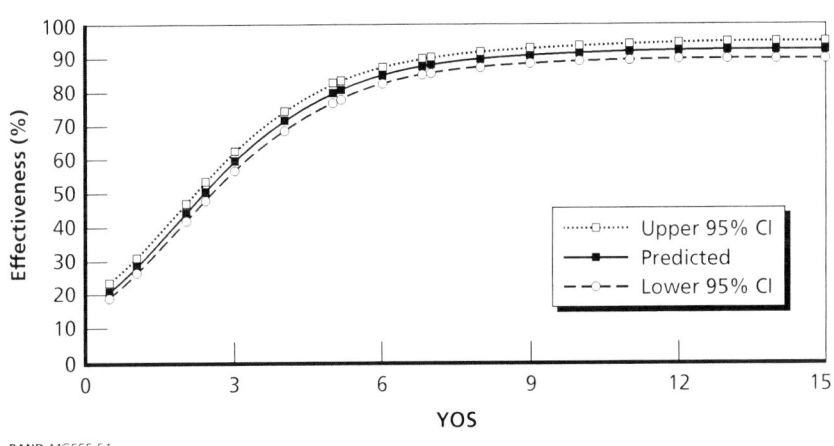

$(\sin(y))_2$ transformation was applied to return the estimate to the original scale. In Figure F.1, the predicted values and their respective 95-percent CIs are plotted for the 2T3X1 AFSC. Figures F.2 through F.7 depict the other six AFSCs. These curves represent the response and confidence interval for the average person in each specialty. Predicted values of effectiveness approach 100-percent, but never reach it. Our initial thought in fitting the logistic growth model was that the curves would level off at 100-percent, but several respondents made comments in which they stated that one never quite reaches 100-percent effectiveness because learning and growth continue throughout an airman's career, and this is thus reflected in the curves.

Table F.1 provides the parameter estimates from the logistic model of the transformed data. Note the small standard errors, and hence the narrow CIs, for these values. Also, the low p-values indicate the significance of all five of these parameters. One can use these parameters to obtain a predicted value for effectiveness on a given year and then back-transform that value to the original scale.

Table F.1
Parameter Estimates for 2T3X1 Production

Parameter	Estimate	Std. Error	Confidence Limit		t-Value	p-value
			Lower 95%	Upper 95%		
B_1	1.292	0.022	1.248	1.336	58.396	<0.001
B_2	2.105	0.085	1.937	2.272	24.893	<0.001
B_3	0.507	0.028	0.452	0.561	18.364	<0.001
σ_u^2	0.032	0.005	0.022	0.042	6.405	<0.001
σ_ε^2	0.024	0.001	0.022	0.026	19.531	<0.001

In the following figures, we provide the unscaled productivity curves for the remaining specialties examined in the study. These curves extend from the end of IST (near 0 or 1 YOS) to 15 YOS. The 95-percent CIs are also included with the curves. Parameter estimates for the additional six specialties are found in Table E.3.

Scaling the Effectiveness Curves

As noted in Chapter Three, the productivity curves never reach 100 percent. Our survey attempted to determine the number of YOS at which 100-percent effectiveness is achieved by asking the question in two ways. First, we asked, How many YOS are needed to reach 100-percent effectiveness? To this question, we received a range of YOS. Second, we asked, What is the productivity of an airman with various YOS? Surprisingly, we found a great deal of reluctance on the part of a sufficient number of the respondents to specify 100-percent effectiveness of any particular year. This paradox led us to scale the time-based productivity curves to 100 percent at the average time to reach 100-percent effectiveness as given by the respondent. Thus, we took the curves generated by the statistical estimates below and proportionately adjusted each point such that the "go-to" point equaled 100 percent.

Figures F.2 through F.7 show the original estimates to the productivity curves prior to scaling.

Figure F.2
Plot of Back-Transformed Predicted Values and CIs, AFSC 1A8X1

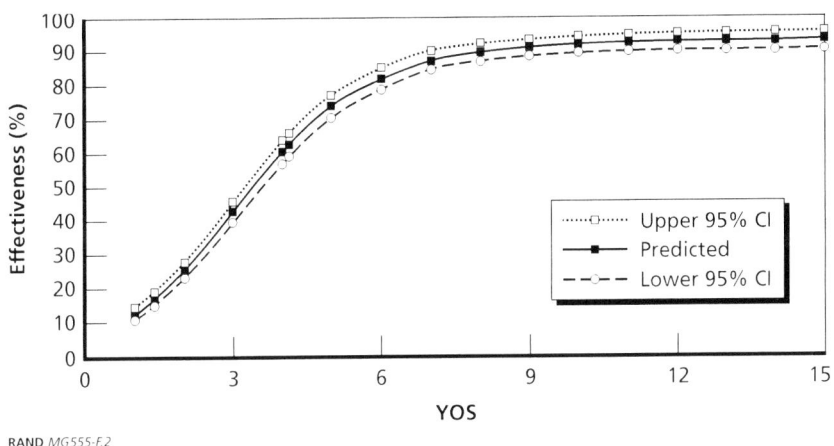

Figure F.3
Plot of Back-Transformed Predicted Values and CIs, AFSC 1N3XX

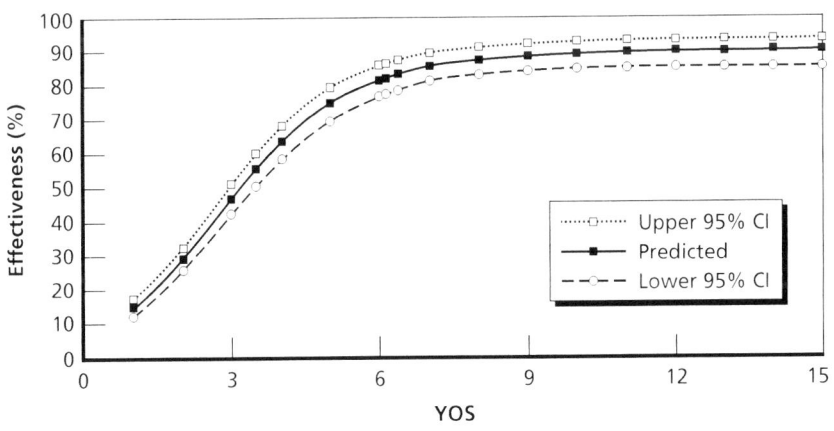

Figure F.4
Plot of Back-Transformed Predicted Values and CIs, AFSC 2A3X3A

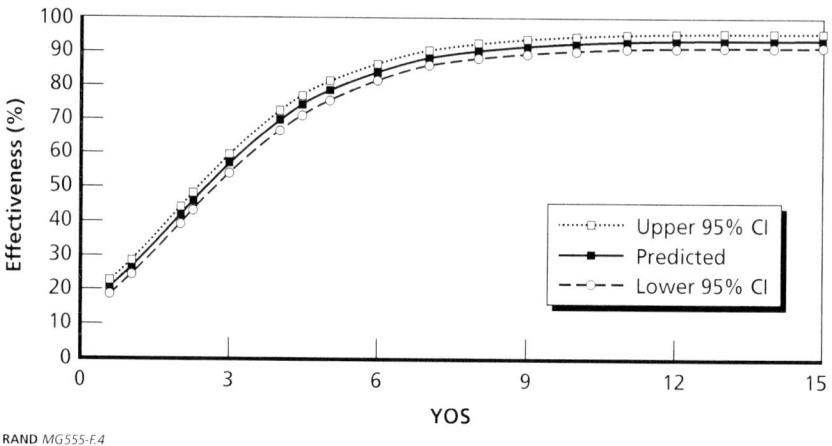

Figure F.5
Plot of Back-Transformed Predicted Values and CIs, AFSC 2A5X1E

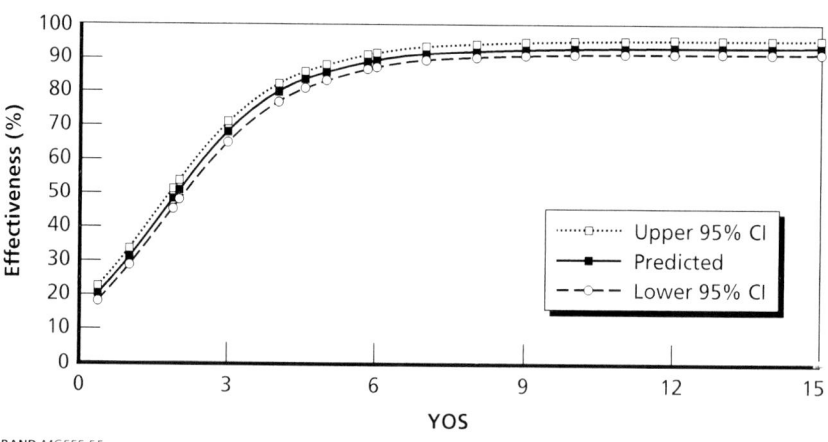

Figure F.6
Plot of Back-Transformed Predicted Values and CIs, AFSC 3P0X1

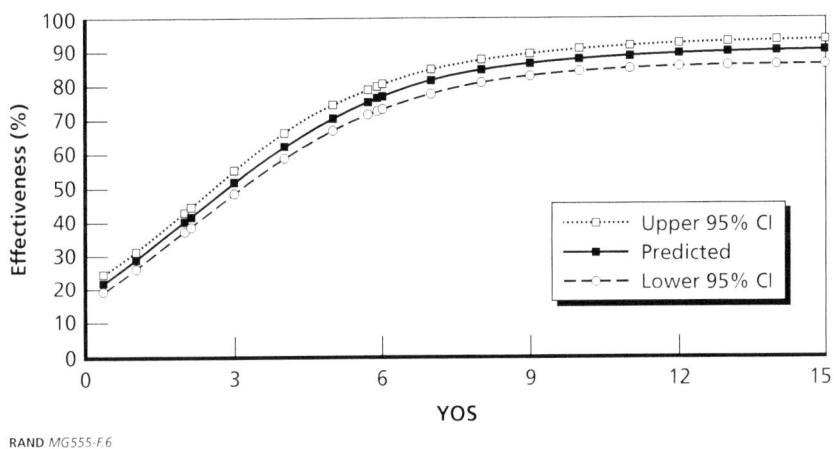

Figure F.7
Plot of Back-Transformed Predicted Values and CIs, AFSC 3E7X1

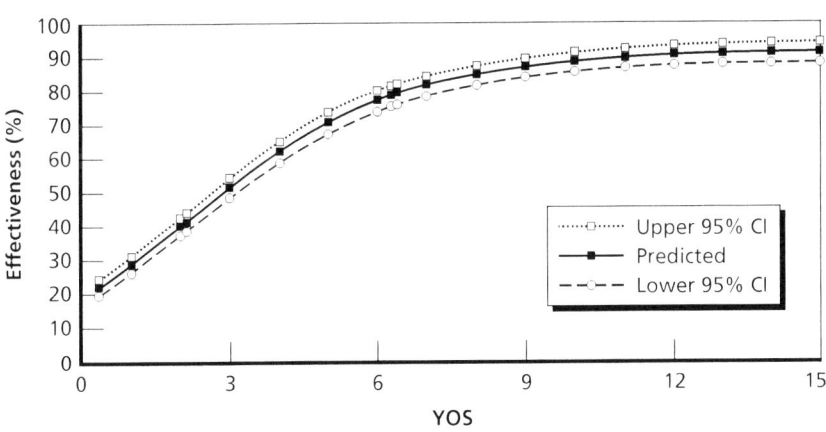

Appendix G

Comments on Adding/Deleting Content from IST

In this appendix, we report on the individual responses given for adding and dropping training blocks. The appendix does not examine the numerical portions of questions 19 and 20 that requested respondents to enter the length of time required to train a task in TT and OJT and estimated productivity change corresponding to the changed training content. Rather, it focuses on only the free response question. This part of the survey gave respondents the opportunity to select the blocks of training that they felt would be most suitable for either addition to or deletion from their specialty's technical training course(s). The purpose of this appendix is to give consideration to all the thoughtfully entered comments, regardless of whether or not the entered numerical responses were logical.

1A8X1 Written Comments

From the overall analysis, there is some indication that the 1A8X1 technical training course has the potential to be shortened. Further analysis of the content of repeated comments provides additional insights into specific recommended potential changes. The fact that 17 separate individuals suggested deleting some (or all) of the technical training at Goodfellow AFB is extremely noteworthy. In a free-response survey, that many respondents making exactly the same recommendation is a very strong indication that those in the field perceive redundancy within the 1A8X1 training pipeline.

Specific suggestions for curriculum additions included the following:

- enhanced weapon system training (8 respondents)
- more hands-on/operational training (7)
- enhanced language training (5)
- cryptologic skill training (3).

Specific suggestions for curriculum deletions included the following:

- Goodfellow AFB (17)
- Enlisted Aircrew Undergraduate Course (11)
- reduce the lag time between courses and waiting for opportunities for flights (5).

1N3XX Written Comments

Overall, the response rate for the 1N3XXs was low, and an unusually low percentage of these individuals chose to respond to this question. Perhaps concerns about the confidentiality requirements of technical school contributed to the low rate. Given that, the 1N3XX respondents are relatively satisfied with the technical training pipeline. Based on the comments, the main focus of improvement efforts was on the language portion of the training, ensuring that it is as comprehensive and up-to-date as possible. All the additions suggest enhancements to the language-training program. The individuals who suggested deleting portions of the language training seemed to indicate that the training is ineffective rather than unnecessary. It is also worth examining the training occurring at Goodfellow AFB, ensuring that unnecessary redundancy is avoided.

Specific suggestions for curriculum additions included the following:

- dialect training (3)
- colloquial/conversational training (3)

- military jargon (2)
- language immersion (2)
- translation skills (1).

Specific suggestions for curriculum deletions included the following:

- Goodfellow AFB (7)
- language training (3).

2A3X3A Written Comments

Both the response patterns and the comments for this specialty were difficult to interpret. The one characteristic that seems to be consistent throughout the written suggestions is the observation that trainees are not arriving on station as ready as they should be. Some of the respondents address this problem by suggesting more time in technical training, so that trainees have the time and opportunity to reach the expected level of proficiency. Other respondents address the problem by suggesting that trainees be put on station sooner. The assumption is that they will start with a lower level of proficiency but that the hands-on OJT will get them up to speed more quickly, and with less redundancy, than the current pipeline training program. As a result, there were a variety of different repeated suggestions for addition and a combination of skill-specific suggestions for deletion along with some drastic recommendations for cutting IST.

Specific suggestions for curriculum additions included the following:

- increase the length of technical training without changing the requirements (8)
- forms/documentation (8)
- Core Automated Maintenance System (CAMS) (7)
- hands-on time (6)
- training specific to one's assigned aircraft (6)

- flight line operations (4)
- launch/recovery (4)
- pre- and post-flight inspections (3)
- flight-line equipment (1).

Specific suggestions for curriculum deletions included the following:

- hot training (5)
- mission ready technician (4)
- fundamentals (4)
- everything except fundamentals (3)
- all of technical training (3)
- all hands-on training (2)
- aircraft servicing (4)
- launch and recovery (3)
- parts installation/removal (3).

2A5X1E Written Comments

Overall, the individuals in this specialty are more in favor of adding to technical training than deleting from it—although not overwhelmingly so. The suggestions for possible additions varied across specific tasks with an emphasis on increasing the amount of hands-on experience in technical training. The deletions were much more general, with the majority of those responding indicating that entire courses in the training sequence were unhelpful or unnecessary. Five respondents used this opportunity to emphasize the need for enhanced training in "respect, followership, and/or motivation."

Specific suggestions for curriculum additions included the following:

- additional hands-on training (12) especially in the form of requests for an actual B1-B (6)
- increased emphasis on the basics (8)

- more emphasis on one's specifically assigned aircraft (7)
- increasing the length of FTD (7)
- troubleshooting (6)
- forms documentation (5)
- CAMS (5).

Specific suggestions for curriculum deletions included the following:

- aircraft fundamentals (8)
- training not related to one's specifically assigned aircraft (8)
- sending trainees directly from the fundamentals course into FTD (6).

2T3X1 Written Comments

The response patterns as well as the comments from this specialty indicate that only additions should be made to technical training. The 2T3X1 specialty had the highest number of respondents who were completely satisfied (12) and an unusually low number of respondents who provided deletion suggestions. Interestingly, no more than two people suggested the same deletion. Instead, the respondents strongly indicated a need for increased troubleshooting training, and suggested a wide variety of tasks that could use more emphasis in the technical training course.

Specific suggestions for curriculum additions included the following:

- troubleshooting training (19)
- equipment familiarization (6)
- engines (6)
- electrical/computer analysis (4)
- scheduled maintenance (4).

Specific suggestions for curriculum deletions included the following:

- scheduled maintenance (2)
- snow removal (2)
- tool identification (2)
- hazardous material (HAZMAT) (2)
- graders (2).

3E7X1 Written Comments

The firefighting course is unique because it trains to a civilian-recognized level of proficiency. This might explain why such a large percentage of the respondents feel that nothing can be deleted but also why the written comments indicate that deployment preparation is lacking within technical training. One interesting source of disagreement among the respondents was on the necessity of medical training (EMT/First Responder) during technical school—similar numbers of respondents suggested adding to it and deleting from it. This could be an indication that the medical training in technical school is incomplete, and respondents are mixed in their opinions of whether to eliminate it completely or enhance it. Finally, five individuals took this opportunity to write in comments specifically addressing the need for additional "customs and courtesies" training—which was unique to this particular specialty.

Specific suggestions for curriculum additions included the following:

- deployment preparation (33)
- medical training (14)
- vehicle operations (7)

- pump training (3)
- ARFF training (3).

Specific suggestions for curriculum deletions included the following:

- HAZMAT training (8)
- medical training (8)
- FACC training (5).

3P0X1 Written Comments

As illustrated in Figure 3.7, the security forces specialty shows the lowest overall satisfaction level of the seven specialties in this study. Although the survey instrument was not designed specifically to measure overall career field satisfaction, the responses did provide support for one possible explanation. Within the comment blocks, a total of 11 individuals voiced concern for merging law enforcement and security, thus making the 3P0X1 specialty "too broad." This theme permeated the addition/deletion suggestions; of the 24 individuals who had specific suggestions for both an addition and a deletion, half implied a trade between two of three broad categories—law enforcement, flightline security, and/or combat operations. In addition, four respondents requested an increase to the length of technical training without any additional training requirements. Given the stresses of the current operational environment, including a shortage of on-station trainers available for OJT, respondents seemed to be emphasizing the need to use IST to get trainees as prepared as possible for the specific jobs that they will be expected to do.

Specific suggestions for curriculum additions included the following:

- increased combat/deployment preparation (21)
- law enforcement (17)
- security training (9).

Specific suggestions for curriculum deletions included the following:

- law enforcement (9)
- rifle fighting techniques (3)
- security (5)
- ground combat skills (4).

General Conclusions

Generating strong conclusions from open-ended survey questions can be extremely difficult, but the suggestions given provide additional support for the recommendations made in this analysis. This is especially true in instances such as the multiple recommendations for adding "troubleshooting" for the 2T3X1s. Such responses are particularly compelling since a large number of respondents gave the same answer without prompting. It is certainly plausible that if all the respondents were given the suggestion of "troubleshooting" as an option for something to add, an even greater percentage would have been in agreement. Similar agreement can be imagined for all the frequently repeated suggestions within these open-ended response blocks.

Cost-Productivity Results for Seven AFSCs

This appendix provides the final cost-effectiveness charts for each of the specialties. These charts provide a quantitative justification for the overall recommendations we made for each of the specialties.

Figure H.1
Airborne Cryptologist (AFSC 1A8X1), Steady State Results

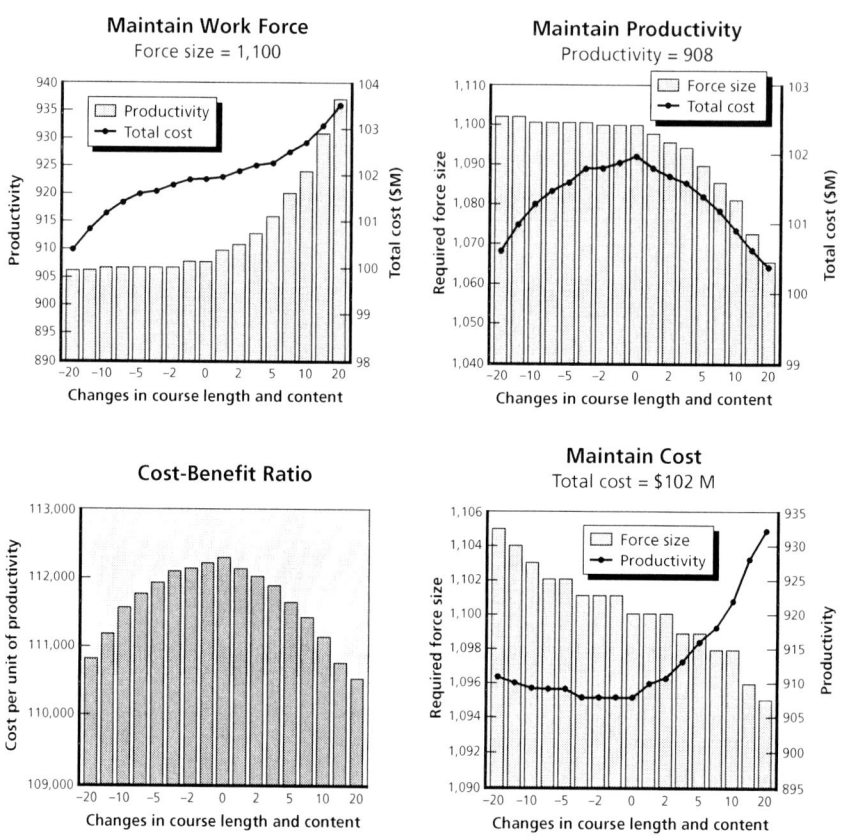

NOTE: Differences must be greater than $1,700 per unit of productivity for statistical significance.

**Figure H.2
Cryptologic Linguist (AFSC 1N3XX), Steady State Results**

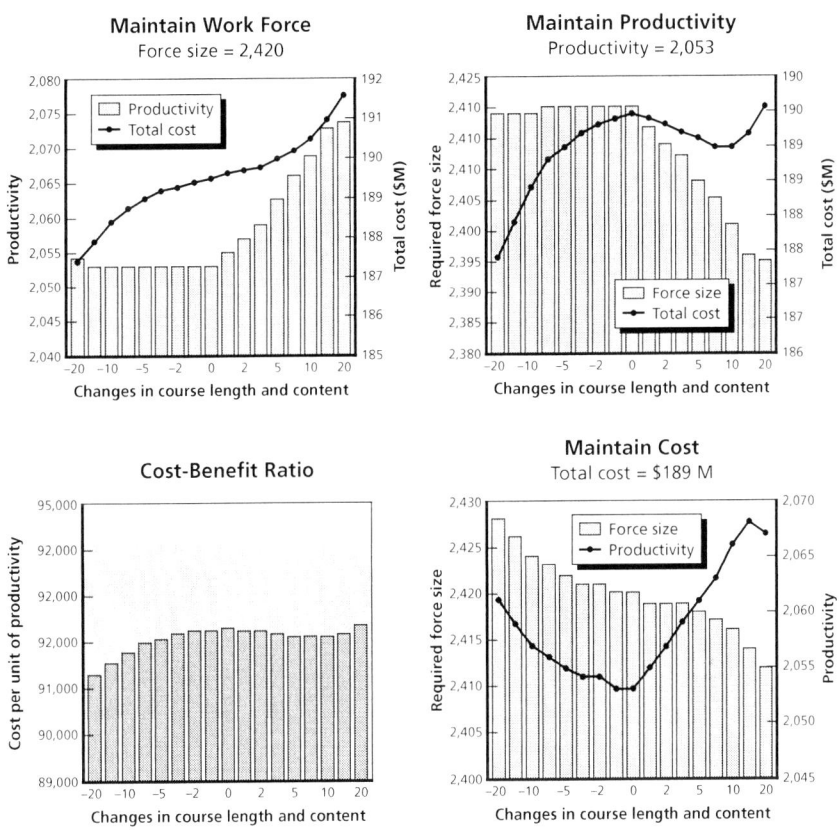

NOTE: Differences must be greater than $2,000 per unit of productivity for statistical significance.

Figure H.3
F-15 Maintenance (AFSC 2A3X3A), Steady State Results

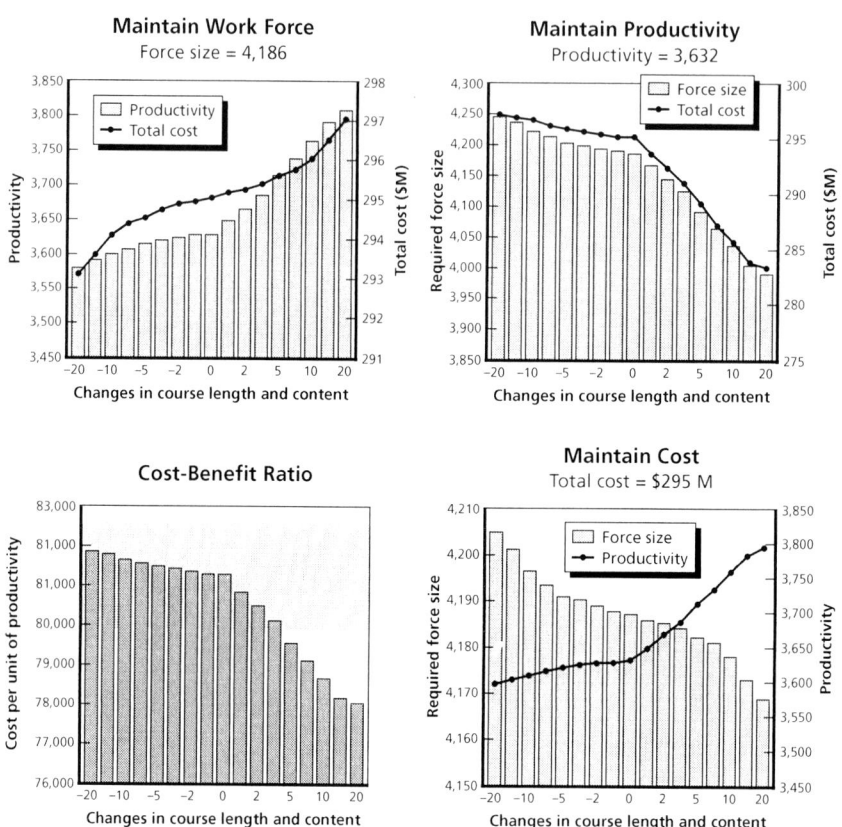

NOTE: Differences must be greater than $1,300 per unit of productivity for statistical significance.

Figure H.4
B-1/B-2 Maintenance (AFSC 2A5X1E), Steady State Results

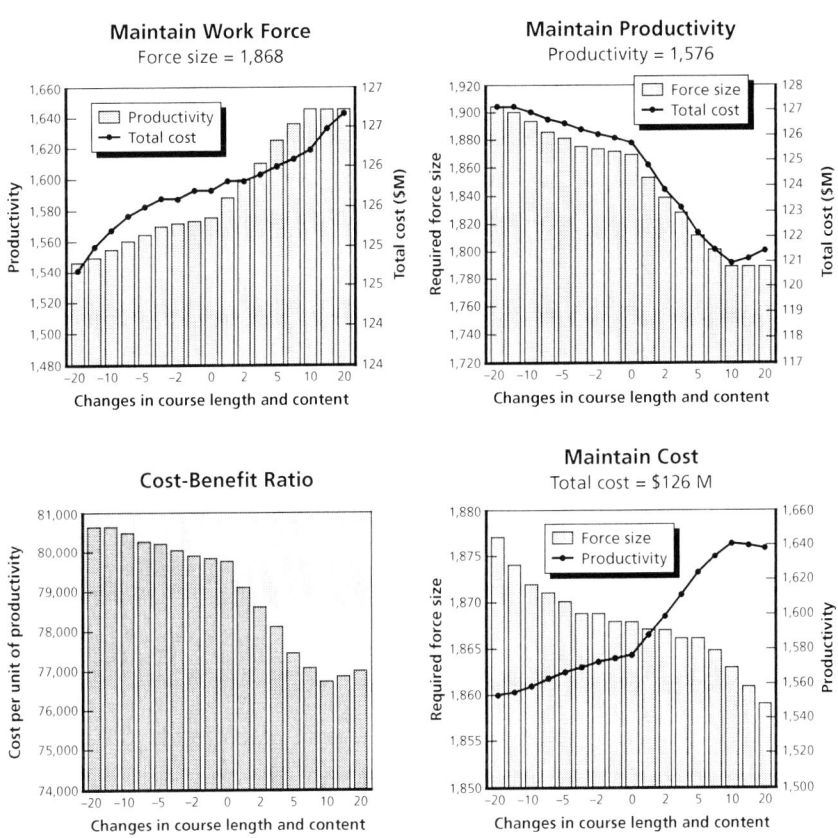

NOTE: Differences must be greater than $1,500 per unit of productivity for statistical significance.

Figure H.5
Special Purpose Vehicle Maintenance (AFSC 2T3X1), Steady State Results

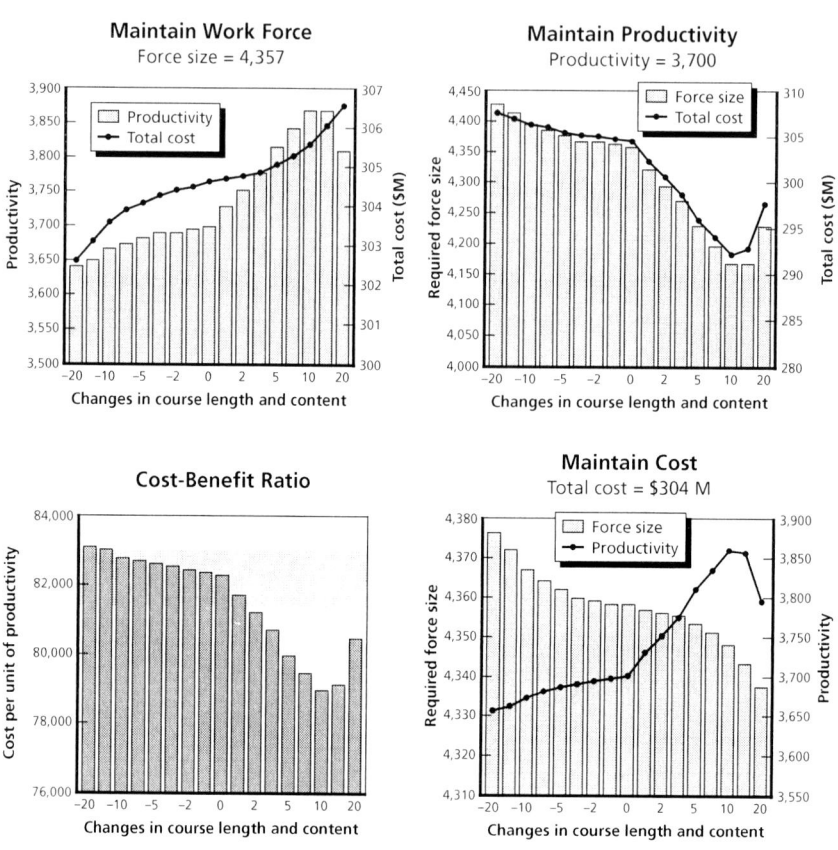

NOTE: Differences must be greater than $1,350 per unit of productivity for statistical significance.

Figure H.6
Fire Fighter (AFSC 3E7X1), Steady State Results

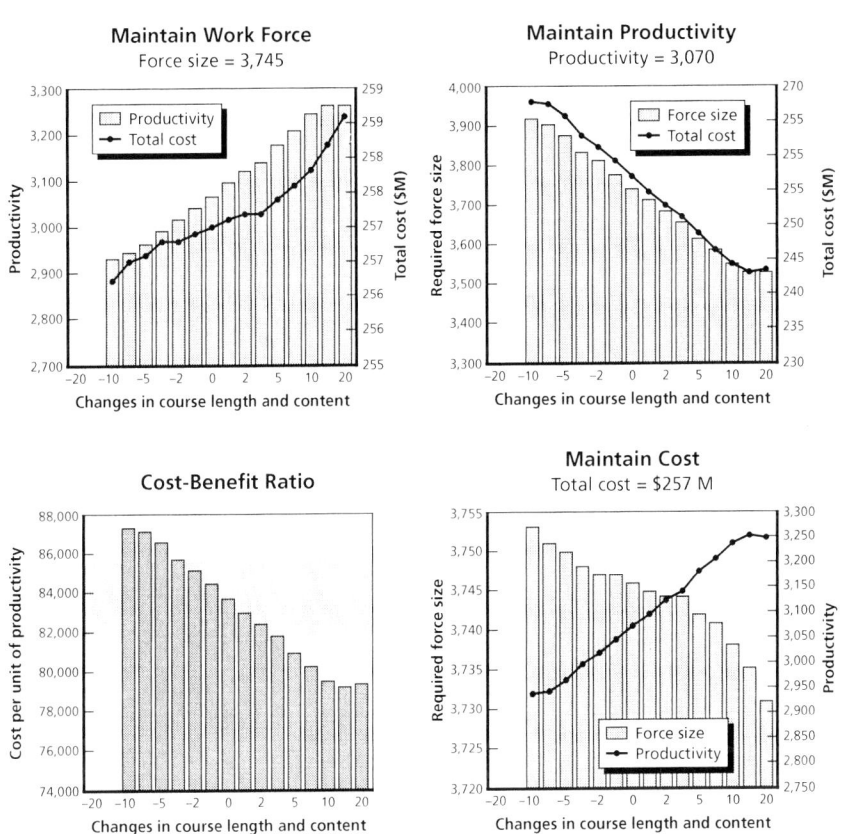

NOTE: Differences must be greater than $2,000 per unit of productivity for statistical significance.

RAND MG555-H.6

Figure H.7
Security Forces (AFSC 3P0X1), Steady State Results

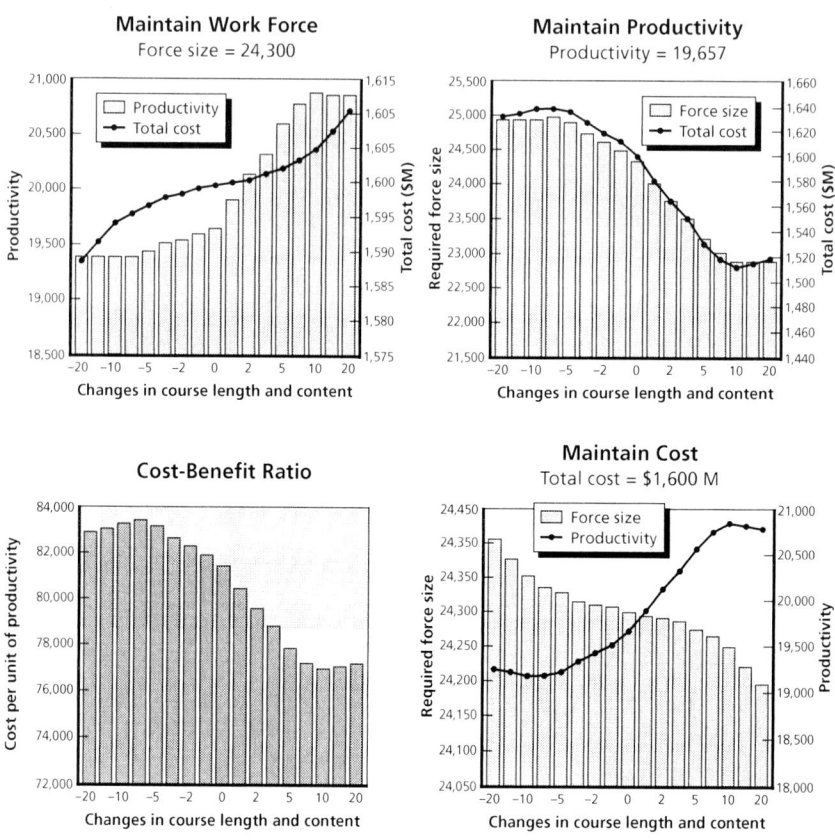

NOTE: Differences must be greater than $2,300 per unit of productivity for statistical significance.

Bibliography

AFI—See Secretary of the Air Force, 1994a and 1994b.

Albrecht, Mark J., *Labor Substitution in the Military Environment: Implications for Enlisted Force Management,* Santa Monica, Calif.: RAND Corporation, R-2330-MRAL, November 1979. As of January 4, 2007: http://www.rand.org/pubs/reports/2005/R2330.pdf

Bennett, Winson R., and Arthur Winford, *Factors That Influence the Effectiveness of Training in Organizations: A Review and Meta-Analysis,* Mesa, Ariz.: Air Force Research Laboratory, Warfighter Training Research Division, AFRL-HE-AZ-TR-2000-0126, June 2001.

Black, Doris, and Robert A. Bottenberg, *Comparison of Technical School and On-the-Job Training as Methods of Skill Upgrading,* Lackland AFB, Tex.: Air Force Human Resources Laboratory, Personnel Division, AFHRL-TR-70-48, December 1970.

Boynton, R. E., N. E. Sciden, and L. E. Vaughn, *NAESU Management of Engineering and Technical Services,* Monterey, Calif.: Naval Postgraduate School, NPS-64-95-001, March 1995.

Broduer, Edmund R., and Karen W. Currie, *Assessment of Initial Technical Training for USAF Supply Officers,* Wright-Patterson AFB, Ohio: Air Force Institute of Technology, AFIT/GLM/LSM/845-5, September 1984.

Carpenter-Huffman, Polly, *Putting On-the-Job Training Under New Management to Improve Its Effectiveness,* Santa Monica, Calif.: RAND Corporation, P-6314, March 1979.

Carpenter-Huffman, Polly, *The Cost-Effectiveness of On-the-Job Training,* Santa Monica, Calif.: RAND Corporation, P-6451, February 1980.

Christal, Raymond, *Development of Task-Level Job Performance Criteria,* Lackland AFB, Tex.: Air Force Human Resources Laboratory, October 21, 1969.

Defense Education and Training Executive Committee, *Report on the OJT StudyTask: On-the-Job Training in the Department of Defense,* January 1981.

Flamholtz, Eric G., *Human Resource Accounting: Advances in Concepts, Methods and Applications,* Boston, Mass.: Kluwer Academic Publishing, 1999.

Fleming, Kenneth H., James Cowardin, Kenneth Reynolds, and David Nielsen, *A Methodology for Estimating the Full Cost of Replacing Trained Air Force Personnel,* Colorado Springs, Colo.: Department of Economics and Geography, United States Air Force Academy, USAFA-TR-87-1, January 1987.

Flournoy, Daniel B., *An Evaluation of the Air Force Specialty Training Standard,* Maxwell AFB, Ala.: Air Command and Staff College, 0755-78, April 1978.

Foxon, Marguerite, "Evaluation of Training and Development Programs: A Review of the Literature," *Australian Journal of Educational Technology,* Volume 5, Number 2, 89-104, 1989. As of January 4, 2007:
http://www.ascilite.org.au/ajet/ajet5/foxon.html

Gay, Robert M., *Estimating the Cost of On-the-Job Training in Military Occupations: A Methodology and Pilot Study,* Santa Monica, Calif.: RAND Corporation, R-1351-ARPA, April 1974. As of January 4, 2007:
http://www.rand.org/pubs/reports/2006/R1351.pdf

Gay, Robert M. and Mark J. Albrecht, *Specialty Training and the Performance of First-Term Enlisted Personnel,* Santa Monica, Calif.: RAND Corporation, R-2191-ARPA, April 1979. As of January 4, 2007:
http://www.rand.org/pubs/reports/2006/R2191.pdf

Gay, Robert M. and Gary R. Nelson, *Cost and Efficiency in Military Specialty Training,* Santa Monica, Calif.: RAND Corporation, P-5160, January 1974.

Gould, R. Bruce, *Reported Job Interest and Perceived Utilization of Talents and Training by Airmen in 97 Career Ladders,* Lackland AFB, Tex.: Air Force Human Resources Laboratory, Personnel Research Division, AFHRL-72-7, January 1972.

Haggstrom, Gus W., Winston K. Chow, and Robert M. Gay, *Productivity Profiles of First-Term Enlisted Personnel,* Santa Monica, Calif.: RAND Corporation, N-2059-RC, February 1984. As of January 4, 2007:
http://www.rand.org/pubs/notes/2005/N2059.pdf

Hanser, Lawrence M., Joyce N. Davidson, and Cathleen Stasz, *Who Should Train? Substituting Civilian-Provided Training for Military Training,* Santa Monica, Calif.: RAND Corporation, R-4119-FMP, 1991.

Hostage, Brig Gen Gilmary, Commander, Air Education and Training Command, *Training Evaluation,* AETCI 36-2201, January 24, 2005.

Horowitz, Stanley A. and Bruce N. Angier, *Costs and Benefits of Training and Experience,* Alexandria, Va.: Center for Naval Analyses, Professional Paper No. 425, January 1985.

Hosek, James R., and Christine E. Peterson, *Developing an Initial Skill Training Database: Rationale and Content,* Santa Monica, Calif.: RAND Corporation, N-2675-FMP, November 1988.

Kavanagh, Jennifer, *Determinants of Productivity for Military Personnel,* Santa Monica, Calif.: RAND Corporation, TR-193-OSD, 2005. As of January 4, 2007: http://www.rand.org/pubs/technical_reports/2005/RAND_TR193.pdf

Main, Ray E., Josephine M. Randel, Barbara A. Morris, *Measuring Training Productivity in Navy Schools,* San Diego, Calif.: Navy Personnel Research and Development Center, TR-92-1, October 1991.

Manacapilli, Thomas, Bart Bennet, Lionel Galway, and Joshua Weed, *Air Education and Training Command Cost and Capacity System: Implications for Organizational Change,* Santa Monica, Calif.: RAND Corporation, MR-1797-AF, 2004. As of January 4, 2007: http://www.rand.org/pubs/monograph_reports/2005/MR1797.pdf

Montgomery, D. C., *Design and Analysis of Experiments,* 5th ed., New York: John Wiley & Sons, 2000.

Oliver, S. A., J. Ausink, T. Manacapilli, J. Drew, S. Naylor, C. Boone, *An Analysis of the Cost and Valuation of Air Force Aircraft Maintenance Personnel,* Maxwell AFB, Ala.: Air Force Logistics Management Agency, LM200107900, 2002.

Quester, Aline, and Alan Marcus, *An Evaluation of the Effectiveness of Classroom and On-the-Job Training,* Alexandria, Va.: Center for Naval Analyses, CRM 86-26, February 1986.

Ruck, Hendrick, Nancy Thompson, and David Thomson, *The Collection and Prediction of Training Emphasis Ratings for Curriculum Development,* Brooks AFB, Tex.: Air Force Human Resources Laboratory, October 1978.

———, *Task Training Emphasis For Determining Training Priority,* Brooks AFB, Tex.: Air Force Human Resources Laboratory, Manpower and Personnel Division, AFHRL-TP-86-85, August 1987.

Secretary of the Air Force, *Enlisted Classification,* Air Force Manual 36-2108, October 1994a. As of October 23, 2006: http://www.e-publishing.af.mil/pubfiles/af/36/afman36-2108/afman36-2108.pdf

———, *U.S. Air Force Cost and Planning Factors,* AFI 65-503, February 1994b. As of October 17, 2006: http://emissary.acq.osd.mil/inst/share.nsf/ 978C9ED6B029DC5C85256FA10045B162/$FILE/afi65-503.pdf

———, *Air Force Training Program: Career Field Education and Training,* AFI 36-2201, Vol. 5, June 2004. As of October 18, 2006: http://www.e-publishing.af.mil/pubfiles/af/36/afi36-2201v5/afi36-2201v5.pdf

Smith, Keith E., *Cost Comparison of Technical Training School Versus Unit Training Methods for Direct-Duty Airmen in Civil Engineering Specialty Codes,* Wright-Patterson AFB, Ohio: Air Force Institute of Technology, AFIT/GLM/LSH/865-24, September 1986.

Sokal, Robert R., and F. James Rohlf, *Biometry: The Principles and Practice of Statistics in Biological Research,* 3rd ed., New York: W. H. Freeman and Company, 1995.

Stephenson, Robert W., and James R. Burkett, *On-the-Job Training in the Air Force: A Systems Analysis,* Lowry AFB, Colo.: Air Force Human Resources Laboratory, Technical Training Division, AFHRL-TR-75-83, December 1975.

Stone, Brice, Jonathan Fast, and Bruce Dunson, *Unit Task-Based Data to Establish Relationships Between Training Time and Performance: Tradeoffs,* San Antonio, Tex.: Metrica, Inc. As of January 4, 2007: http://www.ijoa.org/papers/ioaw99/html/stone.html

Warm, Ronnie, J. Thomas Roth, and Jean A. Fitzpatrick, *On-the-Job Training: Development and Assessment of a Methodology for Generating Task Proficiency Evaluation Instruments,* Lowry AFB, Colo.: Air Force Human Resources Laboratory, Training Systems Division, AFHRL-TR-86-22, September 1986.

Weiher, Rodney, and Stanley Horowitz, *The Relative Costs of Formal and On-the-Job Training for Navy Enlisted Occupations,* Arlington, Va.: Center for Naval Analyses, Professional Paper No. 83, November 1971.

Weissmuller, Johnny J., and J. L. Mitchell, *Self-Prioritized Inventory Administration to Maximize Validity,* San Antonio, Tex.: Metrica, Inc., 1998.

Weissmuller, Johnny J., Martin Dittmar, and R. A. Moon, "A Validation of Automated Training Indicators (ATIs)," *Proceedings of the 8th International Occupational Analysts Workshop,* San Antonio, Tex., June 1993, pp. 306–310.

Wilson, Thurlow R., Joseph A. Olmstead, and Robert C. Trexler, *On-the-Job Training and Social Learning Theory: A Literature Review,* Alexandria, Virginia: Human Resources Research Organization, SR-ETSD-80-5, May 1980.

Zsambok, Caroline E., George L. Kaemph, Beth Crandall, and Molly Kyne, *A Comprehensive Program to Deliver On-the-Job Training,* Fairborn, Ohio: Klein Associates Inc., ARI Contractor Report 97-18, July 1997.